T0297618

Op Amps for Everyone

Op Amps for Everyone

Fourth Edition

Bruce Carter

AMSTERDAM • BOSTON • HEIDELBERG • LONDON
NEW YORK • OXFORD • PARIS • SAN DIEGO
SAN FRANCISCO • SINGAPORE • SYDNEY • TOKYO
Newnes is an imprint of Elsevier

Newnes is an imprint of Elsevier
The Boulevard, Langford Lane, Kidlington, Oxford, OX5 1GB, UK
225 Wyman Street, Waltham, MA 02451, USA

Second edition 2003
Third edition 2009
Fourth edition 2013

Notices

Knowledge and best practice in this field are constantly changing. As new research and experience broaden our
understanding, changes in research methods, professional practices, or medical treatment may become necessary.

Practitioners and researchers must always rely on their own experience and knowledge in evaluating and using
any information, methods, compounds, or experiments described herein. In using such information or methods
they should be mindful of their own safety and the safety of others, including parties for whom they have a
professional responsibility.

To the fullest extent of the law, neither the Publisher nor the authors, contributors, or editors, assume any liability
for any injury and/or damage to persons or property as a matter of products liability, negligence or otherwise, or
from any use or operation of any methods, products, instructions, or ideas contained in the material herein.

British Library Cataloguing-in-Publication Data
A catalogue record for this book is available from the British Library

Library of Congress Cataloging-in-Publication Data
A catalog record for this book is available from the Library of Congress

ISBN: 978-0-12-391495-8

For information on all Newnes publications
visit our web site at www.newnespress.com

Printed and bound in the United States

13 14 15 16   10 9 8 7 6 5 4 3 2 1

Working together to grow
libraries in developing countries

www.elsevier.com | www.bookaid.org | www.sabre.org

ELSEVIER   BOOK AID
International   Sabre Foundation

Contents

List of Figures

List of Tables

List of Abbreviations

ADC	analog-to-digital converter
AGC	automatic gain control
CFA	current feedback amplifier
DAC	digital-to-analog converter
DSL	digital subscriber line
DSP	digital signal processor
EMI	electromagnetic interference
ESD	electrostatic discharge
ESR	equivalent series resistance
FCC	Federal Communications Commission
FET	field-effect transistor
FSV	full-scale voltage
GBW	gain−bandwidth
GSM	Global System for Mobile Communications
HMP	high melting point
IC	integrated circuit
IF	intermediate frequency
J-FET	junction field-effect transistor
LSB	least significant bit
MFB	multiple feedback
NEMA	National Electrical Manufacturers Association
PCB	printed circuit board
RF	radiofrequency

RFI	radiofrequency interference
RIAA	Recording Industry Association of America
RMS	root mean square
RRI	rail-to-rail input
RRIO	rail-to-rail input/output
RRO	rail-to-rail output
SAW	surface acoustic wave
SNR	signal-to-noise ratio
TTL	transistor–transistor logic
UHF	ultra-high frequency
VFA	voltage feedback amplifier

CHAPTER 1

The Op Amp's Place in the World

1.1 An Unbounded Gain Problem

In 1934 Harry Black [1] commuted from his home in New York City to work at Bell Labs in New Jersey by way of a railroad and ferry. The ferry ride relaxed Harry, enabling him to do some conceptual thinking. Harry had a tough problem to solve; when phone lines were extended long distances, they needed amplifiers, and undependable amplifiers limited phone service. Initial tolerances on the gain were poor, but that problem was solved with an adjustment. Unfortunately, even when an amplifier was adjusted correctly at the factory, the gain drifted so much during field operation that the volume was too low or the incoming speech was distorted.

Many attempts had been made to make a stable amplifier, but temperature changes and power supply voltage extremes experienced on phone lines caused uncontrollable gain drift. Harry knew that passive components had much better drift charactcristics than active components; thus if an amplifier's gain could be made dependent on passive components, the problem would be solved. During one of his ferry trips, Harry's fertile brain conceived a novel solution for the amplifier problem, and he documented the solution while riding on the ferry.

1.2 The Solution

The solution was first to build an amplifier that had much more gain and bandwidth than the application required. Then some of the amplifier output signal could be fed back to the input in a manner that made the circuit gain (circuit is the amplifier and feedback components) dependent on the feedback circuit rather than the amplifier gain. The circuit gain was then dependent on the stable passive feedback components rather than the active amplifier. This is called negative feedback, and it is the underlying operating principle for all modern day operational amplifier circuits. Harry had documented the first intentional feedback circuit during his ferry ride. I am sure unintentional feedback circuits had been built prior to that time, but the designers ignored the effect!

DOI: http://dx.doi.org/10.1016/B978-0-12-391495-8.00001-5

I can hear the squeals of anguish coming from the managers and amplifier designers at Bell Labs. I imagine that they said something like this, "It is hard enough to achieve 30 kHz gain–bandwidth (GBW), and now this fool wants me to design an amplifier with 3 MHz GBW. But, all he is making is an audio amplifier". Nevertheless, they accomplished it, the problem was solved, and the benefits of negative feedback were realized. The operational amplifier was born — even though it was not recognized as a component in its own right, or called an operational amplifier until years later.

There is a minor problem that Harry did not discuss in detail, and that is the oscillation problem. It seems that circuits designed with large open-loop gains sometimes oscillate when the loop is closed. A lot of people investigated the instability effect, and it was pretty well understood in the 1940s, but solving stability problems involved long, tedious, and intricate calculations. Years passed without anybody making the solution simpler or more understandable.

In 1945 H.W. Bode presented a system for analyzing the stability of feedback systems using graphical methods. Until this time, feedback analysis was done by multiplication and division, and calculation of transfer functions was a time-consuming and laborious task. Bode presented a logarithmic technique that transformed the intensely mathematical process of calculating a feedback system's stability into graphical analysis that was simple and perceptive. Feedback system design was no longer an art dominated by a few electrical engineers who were also accomplished mathematicians. Any electrical engineer could use Bode's methods to find the stability of a feedback circuit, so the application of feedback circuits began to grow. Operational amplifiers had become easier to use and understand. There was not much call for operational amplifiers, however, until analog computers and transducers came of age.

1.3 The Birth of the Op Amp as a Component

The first computers were analog computers! Programming consisted of configuring wiring and passive components to a series of circuits that performed mathematical operations on voltages. The heart of the analog computer was Harry's invention: a device called an operational amplifier because it could be configured to perform many mathematical operations such as multiplication, addition, subtraction, division, integration, and differentiation on the input signals. The name was shortened to the familiar *op amp*, as it is now known.

General-purpose analog computers were found in universities and large company laboratories because they were critical to the research work done there. Although early op amps were designed for analog computers, it was soon discovered that op amps had other uses and were very handy to have around the physics lab. There was a parallel requirement for transducer signal conditioning in lab experiments, and op amps found their way into signal conditioning applications. As the signal conditioning applications expanded, the demand for op amps grew beyond the analog computer requirements.

The hard wiring limitation eventually caused the declining popularity of the analog computer. Even when analog computers lost favor to digital computers, the op amp survived because of its importance in universal analog applications. Eventually digital computers completely replaced analog computers (a sad day for real-time measurements), but the demand for op amps increased as measurement applications increased.

1.3.1 The Vacuum Tube Era

The first signal conditioning op amps were constructed with vacuum tubes before the introduction of transistors, so they were large and bulky. During the 1950s, miniature vacuum tubes that worked from lower voltage power supplies enabled the manufacture of op amps shrunk to the size of a brick used in house construction, so the op amp modules were nicknamed *bricks*. Vacuum tube size and component size decreased until an op amp was shrunk to the size of a single octal vacuum tube.

One of the first commercially available op amps was the model K2-W, sold by George A. Philbrick Research. It consisted of two vacuum tubes, and operated from a ± 300 V power supply! If that is not enough to make you cringe, then its fully differential nature is sure to. A fully differential op amp, as opposed to the more familiar single-ended op amp, has two outputs: a non-inverting output and an inverting output. It requires you to close two feedback paths, not just one. Before panic sets in, the two feedback pathways only require duplication of components, not an entirely new design methodology. Fully differential op amps are currently enjoying a resurgence because they are ideal components for driving the inputs of fully differential analog-to-digital converters (ADCs). They also find use in driving differential signal pairs such as digital subscriber line (DSL) and balanced 600Ω audio. Suffice it to say, op amps have come full circle since their original days.

1.3.2 The Transistor Era

Transistors were commercially developed in the 1960s, and they further reduced op amp size to several cubic inches, but the nickname brick still held on. Now this nickname is attached to any electronic module that uses potting compound or non-integrated circuit (IC) packaging methods. Most of these early op amps were made for specific applications, so they were not necessarily general purpose. The early op amps served a specific purpose, but each manufacturer had different specifications and packages; hence, there was little second sourcing among the early op amps.

1.3.3 The Integrated Circuit Era

Although ICs were developed during the late 1950s and early 1960s, it was not until the mid-1960s that Fairchild released the μA709. This was the first commercially successful IC op amp, and Robert J. Widler designed it. The μA709 had its share of problems, but any competent analog engineer could use it, and it served in many different analog applications. The major drawback of the μA709 was stability; it required external compensation. In addition, the μA709 was quite sensitive because it had a habit of self-destructing under any adverse condition. The legacy of its instability remains – few uncompensated amplifiers are sold today owing to the problem of misapplication. Stability remains one of the least understood aspects of op amp design, and one of the easiest ways to misapply an op amp. Even engineers with years of analog design experience have differing opinions on the topic. The wise engineer, however, will look carefully at the op amp data sheet, and not attempt a gain less than its specification. It may be counterintuitive, but the op amp is least stable at its lowest specified gain. Future chapters will delve more deeply into this phenomenon.

The μA741 followed the μA709, and it is an internally compensated op amp that does not require external compensation if operated under data sheet conditions. It is also much more forgiving than the μA709. The legacy of the μA741 is much more positive than its predecessor's. In fact, the part number "741" is etched into the memory of practically every engineer in the world, much like the "2N2222" transistor and the "1N4148" diode. It is usually the first part number that comes to mind whenever an engineer thinks of an op amp. Unlike the μA709, the μA741 will work unless grossly misapplied – a fact that has endeared it to generations of engineers. Its power supply requirements of ± 15 V have given rise to hundreds of power supply components that generate these levels, much as $+5$ V has been driven by transistor–transistor logic (TTL) and ± 12 V has been driven by RS232 serial interfaces. For many years, every op amp introduced used the same ± 15 V power supplies as did the μA741. Even today, the μA741 is an excellent choice where wide dynamic range and ruggedness are required.

There has been a never-ending series of new op amps released each year since the introduction of the μA741, and their performance and reliability have improved to the point where present-day op amps can be used for analog applications by anybody who can understand a data sheet. The latest generations of op amps cover the frequency spectrum from 5 kHz GBW for extremely low power devices to beyond 3 GHz GBW. The supply voltage ranges from guaranteed operation at 0.9 V to absolute maximum voltage ratings of 1000 V. The input current and input offset voltage has fallen so low that customers have problems verifying the specifications during incoming inspection. The op amp has truly become the universal analog IC because it performs all analog tasks. It can function as a line-driver, amplifier, level shifter, oscillator, filter, signal conditioner, actuator driver, current source, and voltage source, and in many other applications.

It should be noted that there is no op amp that is universally applicable. An op amp that is ideal for transducer interfaces will not work at all for radio frequency (RF) applications. An op amp with good RF performance might have miserable DC specifications. The hundreds of op amp models offered by manufacturers are all optimized in slightly different ways, so your task is to weed through those hundreds of devices and find the ones that are appropriate for their application. This edition includes a design methodology for doing so, at least in the case of signal chains.

This book deals with op amp circuits, not with the innards of op amps. It does not get bogged down in detailed calculations. Engineers should not have to be mathematicians to perform routine designs. Math should be required only for advanced applications, not reinvented over and over again for routine ones. Using this book, you can start at the level you understand, and quickly move on to advanced topics as needed.

The op amp will continue to be a vital component of analog design because it is such a fundamental component. Each generation of electronics equipment integrates more functions on silicon and takes more of the analog circuitry inside the IC. Don't fear; as digital applications increase, analog applications also increase because the predominant supply of data and interface applications is in the real world, and the real world is an analog world. Thus, each new generation of electronics equipment creates requirements for new analog circuits; hence, new generations of op amps are required to fulfill these requirements. Analog design and op amp design are fundamental skills that will be required far into the future.

Reference

1. H.S. Black, Stabilized feedback amplifiers, Bell System Technical Journal 13(January) (1934).

CHAPTER 2

Review of Op Amp Basics

2.1 Introduction

Although this book minimizes math, some algebra is germane to the understanding of analog electronics. Math is presented here in the manner in which it is used later. Circuits are a mix of passive and active components. The components are arranged in a manner that enables them to perform some desired function. The resulting arrangement of components is called a circuit or sometimes a circuit configuration.

When the design has progressed to the point that a circuit exists, equations must be written to predict and analyze circuit performance. Textbooks are filled with rigorous methods for equation writing, and this review of circuit theory does not supplant those textbooks. However, a few equations are used so often that they should be memorized, and these equations are considered here.

There are almost as many ways to analyze a circuit as there are electronic engineers, and if the equations are written correctly, all methods yield the same answer. There are some simple ways to analyze the circuit without completing unnecessary calculations, and these methods are illustrated here.

2.2 Basic Concepts

2.2.1 Ohm's Law

Ohm's law is stated in Equation 2.1, and is fundamental to all electronics. Ohm's law can be applied to a single component, to any group of components, or to a complete circuit. When the current flowing through any portion of a circuit is known, the voltage dropped across that portion of the circuit is obtained by multiplying the current times the resistance:

$$V = IR \tag{2.1}$$

In Figure 2.1, the current (I) flows through the total resistance (R), and the voltage (V) is dropped across R.

DOI: http://dx.doi.org/10.1016/B978-0-12-391495-8.00002-7

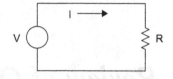

Figure 2.1
Ohm's Law

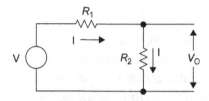

Figure 2.2
Voltage Divider Rule

2.2.2 The Voltage Divider Rule

When the output of a circuit is not loaded, the voltage divider rule can be used to calculate the circuit's output voltage. Assume that the same current flows through all circuit elements (Figure 2.2). Equation 2.2 is written using Ohm's law as $V = I (R_1 + R_2)$. Equation 2.3 is written as Ohm's law across the output resistor.

$$I = \frac{V}{R_1 + R_2} \tag{2.2}$$

$$V_{OUT} = IR_2 \tag{2.3}$$

Substituting Equation 2.2 into Equation 2.3 and using algebraic manipulation yields Equation 2.4:

$$V_{OUT} = V\frac{R_2}{R_1 + R_2} \tag{2.4}$$

A simple way to remember the voltage divider rule is that the output resistor is divided by the total circuit resistance. This fraction is multiplied by the input voltage to obtain the output voltage. Remember that the voltage divider rule always assumes that the output resistor is not loaded; the equation is not valid when the output resistor is loaded by a parallel component. Fortunately, most circuits following a voltage divider are input circuits, and input circuits are usually high-resistance circuits. When a fixed load is in parallel with the output resistor, the equivalent parallel value comprised of the output resistor and loading resistor can be used in the voltage divider calculations with no error. Many people ignore the load resistor if it is 10 times greater than the output resistor value, but this will lead to a 10% error.

2.2.3 Superposition

Superposition is a theorem that can be applied to any linear circuit. Essentially, when there are independent sources, the voltages and currents resulting from each source can be calculated separately, and the results are added algebraically. This simplifies the calculations because it eliminates the need to write a series of loop or node equations. An example is shown in Figure 2.3.

When V_1 is grounded, V_2 forms a voltage divider with R_3 and the parallel combination of R_2 and R_1. The output voltage for this circuit (V_{OUT2}) is calculated with the aid of the voltage divider equation (Equation 2.4). The circuit is shown in Figure 2.4. The voltage divider rule yields the answer quickly.

$$V_{OUT2} = V_2 \frac{R_1 || R_2}{R_3 + R_1 || R_2} \tag{2.5}$$

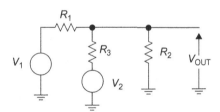

Figure 2.3
Superposition Example

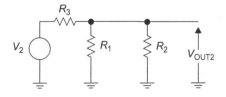

Figure 2.4
When V_1 is Grounded

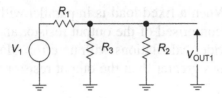

Figure 2.5
When V_2 is Grounded

Likewise, when V_2 is grounded (Figure 2.5), V_1 forms a voltage divider with R_1 and the parallel combination of R_3 and R_2, and the voltage divider theorem is applied again to calculate V_{OUT} (Equation 2.6).

$$V_{OUT1} = V_1 \frac{R_2||R_3}{R_1 + R_2||R_3} \tag{2.6}$$

After the calculations for each source have been made the components are added to obtain the final solution (Equation 2.7):

$$V_{OUT} = V_1 \frac{R_2||R_3}{R_1 + R_2||R_3} + V_2 \frac{R_1||R_2}{R_3 + R_1||R_2} \tag{2.7}$$

Again, the superposition results come out as a simple arrangement that is easy to understand. One looks at the final equation and it is obvious that if the sources are equal and opposite polarity, and when $R_1 = R_3$, then the output voltage is zero.

2.3 Basic Op Amp Circuits

The following circuits are the most commonly used op amp circuits. They assume an ideal op amp component, and therefore will degrade with real-world components (see Section 2.4). Nevertheless, they form a starting point for op amp design and are valid enough if the application is not too critical.

Some quick definitions here:

- Gain: the voltage ratio of op amp stage output voltage to op amp stage input voltage
- Open-loop gain: the gain of the op amp with no feedback
- Feedback: the portion of the op amp output that is fed back, out of phase, with the input signal
- R_f: the "feedback" resistor — connected between the op amp output and inverting input
- R_g: the "gain" resistor — connected between the op amp inverting input and ground.

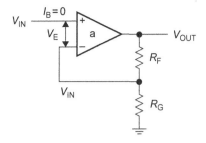

Figure 2.6
The Non-Inverting Op Amp

2.3.1 The Non-Inverting Op Amp

The non-inverting op amp has the input signal connected to its non-inverting input (Figure 2.6), and therefore its input source sees high impedance. There is no input offset voltage because $V_{OS} = V_E = 0$; hence the negative input must be at the same voltage as the positive input. The op amp output drives current into R_f until the negative input is at the voltage V_{IN}. This action causes V_{IN} to appear across R_g.

The voltage divider rule is used to calculate V_{IN}; V_{OUT} is the input to the voltage divider, and V_{IN} is the output of the voltage divider. Since no current can flow into either op amp lead, use of the voltage divider rule is allowed. Equation 2.8 is written with the aid of the voltage divider rule, and algebraic manipulation yields Equation 2.9 in the form of a gain parameter.

$$V_{IN} = V_{OUT} \frac{R_g}{R_g + R_f} \tag{2.8}$$

$$\frac{V_{OUT}}{V_{IN}} = \frac{R_g + R_f}{R_g} = 1 + \frac{R_f}{R_g} \tag{2.9}$$

When R_g becomes very large with respect to R_f, $(R_f/R_g) \Rightarrow 0$ and Equation 2.9 reduces to Equation 2.10:

$$V_{OUT} = 1 \tag{2.10}$$

Under these conditions $V_{OUT} = 1$ and the circuit becomes a unity gain buffer. R_g is usually deleted to achieve the same results, and when R_g is deleted, R_f can be shorted. Some op amps are self-destructive when R_f is shorted, so R_f is used in many buffer designs. When R_f is included in a buffer circuit, its function is to protect the inverting input from an overvoltage to limit the current through the input electrostatic discharge (ESD) structure (typically <1 mA), and it can have almost any value

(20 kΩ is often used). R_f can never be left out of the circuit in a current feedback amplifier design because R_f determines stability in current feedback amplifiers.

2.3.2 The Inverting Op Amp

The non-inverting input of the inverting op amp circuit is grounded as shown in Figure 2.7. One assumption made is that the input error voltage is zero, so the feedback keeps inverting the input of the op amp at a virtual ground (not actual ground but acting like ground). The current flow in the input leads is assumed to be zero; hence the current flowing through R_g equals the current flowing through R_f. Using Kirchhoff's law, we write Equation 2.11; and the minus sign is inserted because this is the inverting input. Algebraic manipulation gives Equation 2.12.

$$I_1 = \frac{V_{IN}}{R_g} = -I_2 = -\frac{V_{OUT}}{R_f} \tag{2.11}$$

$$\frac{V_{OUT}}{V_{IN}} = -\frac{R_f}{R_g} \tag{2.12}$$

Notice that the gain is only a function of the feedback and gain resistors, so the feedback has accomplished its function of making the gain independent of the op amp parameters. The actual resistor values are determined by the impedance levels that you want to establish. If $R_f = 10$ kΩ and $R_g = 10$ kΩ the gain is minus one as shown in Equation 2.12, and if $R_f = 100$ kΩ and $R_g = 100$ kΩ the gain is still minus one. The impedance levels of 10 kΩ or 100 kΩ determine the current drain, the effect of stray capacitance, and a few other points. The impedance level does not set the gain; the ratio of R_f/R_g does.

One final note: the output signal is the input signal amplified and inverted. The circuit input impedance is set by R_g because the inverting input is held at a virtual ground.

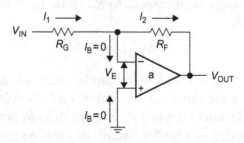

Figure 2.7
The Inverting Op Amp

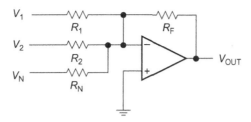

Figure 2.8
The Adder Circuit

2.3.3 The Adder

An adder circuit can be made by connecting more inputs to the inverting op amp (Figure 2.8). The opposite end of the resistor connected to the inverting input is held at virtual ground by the feedback; therefore, adding new inputs does not affect the response of the existing inputs.

Superposition is used to calculate the output voltages resulting from each input, and the output voltages are added algebraically to obtain the total output voltage. Equation 2.13 is the output equation when V_1 and V_2 are grounded. Equations 2.14 and 2.15 are the other superposition equations, and the final result is given in Equation 2.16.

$$V_{OUTN} = -\frac{R_f}{R_N} V_N \tag{2.13}$$

$$V_{OUT1} = -\frac{R_f}{R_1} V_1 \tag{2.14}$$

$$V_{OUT2} = -\frac{R_f}{R_2} V_2 \tag{2.15}$$

$$V_{OUT} = -\left(\frac{R_f}{R_1} V_1 + \frac{R_f}{R_2} V_2 + \frac{R_f}{R_N} V_N \right) \tag{2.16}$$

2.3.4 The Differential Amplifier

The differential amplifier circuit amplifies the difference between signals applied to the inputs (Figure 2.9). Superposition is used to calculate the output voltage resulting from each input voltage, and then the two output voltages are added to arrive at the final output voltage.

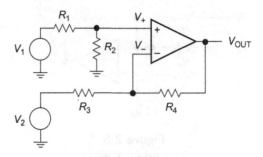

Figure 2.9
The Differential Amplifier

The op amp input voltage resulting from the input source, V_1, is calculated in Equations 2.17 and 2.18. The voltage divider rule is used to calculate the voltage, V_+, and the non-inverting gain equation (Equation 2.18) is used to calculate the non-inverting output voltage, V_{OUT1}.

$$V_+ = V_1 \frac{R_2}{R_1 + R_2} \tag{2.17}$$

$$V_{OUT1} = V_+(G_+) = V_1 \frac{R_2}{R_1 + R_2} \left(\frac{R_3 + R_4}{R_3} \right) \tag{2.18}$$

The inverting gain equation (Equation 2.12) is used to calculate the stage gain for V_{OUT2} in Equation 2.19. These inverting and non-inverting gains are added in Equation 2.20.

$$V_{OUT2} = V_2 \left(\frac{-R_4}{R_3} \right) \tag{2.19}$$

$$V_{OUT} = V_1 \frac{R_2}{R_1 + R_2} \left(\frac{R_3 + R_4}{R_3} \right) - V_2 \frac{R_4}{R_3} \tag{2.20}$$

$$V_{OUT} = (V_1 - V_2) \frac{R_4}{R_3} \tag{2.21}$$

It is now obvious that the differential signal, $(V_1 - V_2)$, is multiplied by the stage gain, so the name differential amplifier suits the circuit. Because it only amplifies the differential portion of the input signal, it rejects the common-mode portion of the input signal.

2.4 Not So Fast!

The above examples are based on an ideal op amp assumption. There is a small, but noticeable error term that stems from the open-loop response of the op amp. By now, you are probably wondering what in the world I am talking about: the inverting gain has always been $-R_f/R_g$ and the non-inverting $1 + R_f/R_g$. Well, yes, and no. If the open-loop response is infinite (the case with an ideal op amp), then the error term is zero. If the open-loop response is large compared to the gain you are trying to achieve, then the error term is so small as to be unnoticeable. But as you increase the stage gain, this error will start to come into play, and the net result is a gain lower than you expect. This is one of those effects that most designers can ignore; they can get away with the simple solution and never see the open-loop effects. It is one of those things like special relativity and the speed of light: the really strange stuff only happens when you get close. Unfortunately, it is a lot easier to get close to the open-loop response of an op amp than it is to approach the speed of light, so you can definitely experience the strange effects of the open-loop response of the op amp.

But I am getting ahead of myself; I need to introduce the open-loop response characteristic of a typical op amp in Figure 2.10.

This plot, adapted from a real data sheet plot, shows the general shape of the open-loop response of a typical op amp. Such plots usually also include a phase response, which I will discuss later when I discuss op amp stability. I have also deleted the units from the frequency plot, as they are not needed for this discussion. The open-loop response could start at 1 Hz for low-speed op amps, or 1 MHz for high-speed. There are six decades of frequency shown on the plot.

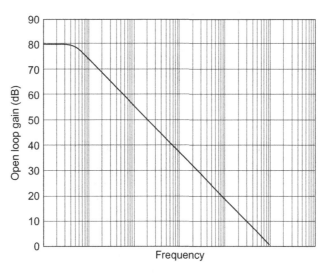

Figure 2.10
Typical Open-Loop Op Amp Response

This plot consists of two main sections: a flat portion and a slanted portion. It is a Bode plot as discussed in Chapter 1, and could be approximated by two straight lines. If you think it looks like the response of a simple low-pass filter, you get extra credit, because that is exactly correct! Most op amps are compensated internally by a single pole loss pass filter in order to swamp out the effects of parasitic elements, a topic for a discussion of stability in Chapter 4. Suffice it to say, the gain you can expect to get out of an op amp stage is strongly related to the frequency of the signal. For example, at the leftmost frequency the open-loop gain is 80 dB, or a gain of 10,000 V/V. So the highest gain you could ever get out of the op amp at that frequency is 10,000. The situation changes dramatically just three decades above. The line slants, and the most gain you could ever get out of the stage is less than 40 dB, or less than 100 V/V. Ouch! That severely limits the stage gain at that frequency. Remember, I said earlier that there is a factor in the open-loop equation that further limits the gain. It allows you to get close to the open-loop response, but never reach it! Let's look at the real equations for the op amp stage gains:

$$\frac{V_{OUT}}{V_{IN}} = \frac{(-a \cdot R_f)/(R_g + R_f)}{1 + (a \cdot R_g)/(R_g + R_f)}, \text{ where } a = \text{open-loop gain}$$

$$\text{If } a \gg \text{ closed-loop gain: } \frac{V_{OUT}}{V_{IN}} = -\frac{R_f}{R_g} \tag{2.22}$$

$$\frac{V_{OUT}}{V_{IN}} = \frac{a}{1 + (a \cdot R_g)/(R_g + R_f)}, \text{ where } a = \text{open-loop gain}$$

$$\text{If } a \gg \text{ closed-loop gain: } \frac{V_{OUT}}{V_{IN}} = 1 + \frac{R_f}{R_g} \tag{2.23}$$

Consider the open-loop response plot above. Assume that the frequency is low, so the open-loop response is 80 dB. Looking at how this affects the actual gains, Table 2.1 shows the inverting case.

Table 2.1: Op Amp Inverting Gains

a	R_g	R_f	Attempted	Actual	Error (%)
10,000	100,000	100,000	−1	−0.9998	−0.0200
10,000	10,000	100,000	−10	−9.9890	−0.1099
10,000	1,000	100,000	−100	−99.0001	−0.9999
10,000	100	100,000	−1,000	−909.0083	−9.0992
10,000	10	100,000	−10,000	−4,999.7500	−50.0025
10,000	1	100,000	−100,000	−9,090.8264	−90.9092
10,000	1	1.00E + 12	−1E + 12	−9,999.9999	−100.0000

Table 2.2: Op Amp Non-Inverting Gains

a	R_g	R_f	Attempted	Actual	Error (%)
10,000	100,000	100,000	2	1.9996	−0.0200
10,000	10,000	100,000	11	10.9879	−0.1099
10,000	1,000	100,000	101	99.9901	−0.9999
10,000	100	100,000	1,001	909.9173	−9.0992
10,000	10	100,000	10,001	5,000.2500	−50.0025
10,000	1	100,000	100,001	9,090.9174	−90.9092
10,000	1	1.00E + 12	1E + 12	9,999.9999	−100.0000

When you attempt a gain of −1 with $R_f = R_g$, the open-loop gain contributes only 0.02% error. If you are using 1% resistors, the error will never be noticed. Even at a gain of −10, the error is only 0.1%, still lost in the resistor tolerance. But at a gain of −100, fully 40 dB below the open-loop response, the error has become 1% and could be noticeable. You can compensate by tweaking the resistors a bit, but by the time a gain of 10,000 is attempted, the error balloons to 50% and no reasonable amount of tweaking will help. If you do something ridiculous like use 1Ω for R_g and 1 TΩ for R_f, the highest gain the stage will provide is still less than −10,000. Just like the speed of light, you can theoretically get as close as you want, but you can never achieve it.

The non-inverting case is similar in Table 2.2: inconsequential errors at low gains, 50% error when you attempt a gain equal to the open-loop gain, and the open-loop gain approached but never reached even with ridiculous extremes of resistors.

This may all sound esoteric and theoretical, far removed from your experience, but keep it in the back of your mind when you are designing gain stages. This is the way op amps really behave, and those simple gain expressions you are used to are only approximations, and they can mislead you.

There are similar expressions for the other simple gain circuits above. Open-loop limitations affect them as well, but the derivations are more complex and lead to similar results. These limitations also affect filter circuits. A very good rule of thumb is to make sure that the open-loop response (Figure 2.10 shows a typical open-loop response) is at least 40 dB above your highest frequency of interest, or errors may occur. In the case of bandpass filters, these limitations will actually force the resonant frequency to shift downward, but that is a topic I will cover in later chapters.

CHAPTER 3

Separating and Managing
AC and DC Gain

3.1 A Small Complication

The previous chapter showed basic op amp gain circuits. There is another aspect to op amp design, one which complicates things. Op amps will amplify DC voltage just as readily as AC voltage. Low-level instrumentation circuits generate their own slight DC offset voltage, which is also amplified by the gain of the circuit. Op amp circuits designed to amplify only AC signals can be constructed with coupling capacitors to block any unwanted DC voltage from being amplified in subsequent stages, which is commonly done in high-gain signal paths. However, this is not acceptable in circuits that must provide the proper DC as well as AC levels.

3.2 Single Supply versus Dual Supply

Closely related to the subject of DC gain due to op amp offsets is the subject of how to power op amps from a single power supply rail, thus intentionally introducing a DC voltage to the op amp inputs. The previous chapter assumed that all op amps were powered from dual or split supplies, which is not the case in the world of portable, battery-powered equipment. When op amps are powered from dual supplies (Figure 3.1), the supplies are normally equal in magnitude and opposed in polarity, and the center tap of the supplies is connected to ground. Any input sources connected to ground are automatically referenced to the center of the supply voltage, so the output voltage is automatically referenced to ground.

When a circuit is powered from a single supply, you have to take into account the DC gain of the op amp as well as the AC gain. You must develop a mindset of determining the DC operating point of the system, which is, of necessity, raised to a potential above ground. Single-supply systems do not have the convenient ground reference that dual-supply systems have; thus a voltage reference of some sort must be employed to ensure that the output voltage swings between the supply voltage and ground, with the midpoint of the voltage swing being the level of the voltage

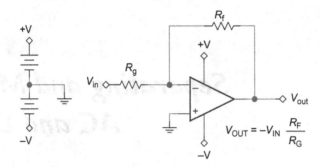

Figure 3.1
Split-Supply Op Amp Circuit

$$V_{OUT} = -V_{IN}\ \frac{R_F}{R_G}$$

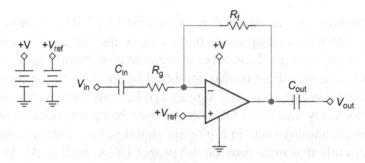

Figure 3.2
Single-Supply Inverting Op Amp Circuit

reference. Input sources connected to ground are actually connected to the negative supply rail in single-supply systems. The voltage reference is then analogous to ground in a dual/split supply system. This requirement for having a voltage reference complicates stage design, because the op amp will dutifully amplify the reference voltage right along with the AC signal voltage. If you are not aware of this, you will quickly discover specifications known as V_{OH} and V_{OL}: the voltage swing limitations of the op amp itself.

Figure 3.2 shows an effective way of combating the problems caused by DC gain on the voltage reference: using coupling capacitors on the input and output of the stage. So how does this work? Assuming the AC input frequency is unaffected by the input and output capacitors (the capacitors are large enough to pass the AC signal unimpeded), the gain of the stage is $-R_f/R_g$, as expected. Since the non-inverting input of the op amp is at $+V_{REF}$, the inverting input will also be at $+V_{REF}$. The DC gain of the stage on $+V_{REF}$ is $1 + R_f/R_g$, but as far as DC is concerned, R_g is infinite. Therefore, the DC gain of the stage reduces to 1, and $+V_{REF}$ appears at the op amp output as well. There are some trade-offs, though:

- This technique will only work for AC signals; if you need to amplify DC voltage as well, you will have to use techniques presented later in this chapter.
- Two power supplies are still needed. What has been gained, though, is that the reference is positive, and therefore can be easily derived from the positive power supply rail, or may already be present in the system (from an analog-to-digital converter, for example).
- As alluded to above, the input capacitor also forms a high-pass filter with a -3 dB point of $1/2\pi R_g C_{IN}$. It is your responsibility to make sure this high-pass breakpoint does not affect the signal you are interested in amplifying.
- The output capacitor also makes a high-pass filter with the input resistance of the subsequent stage, or subsequent circuit. This characteristic also has to be taken into account.

For the non-inverting case, the situation becomes just a bit more complicated. Figure 3.3 shows the modifications required to operate a non-inverting gain stage off a single power supply. Looking at the way this circuit operates, the AC gain on the input voltage is $1 + R_f/R_g$, with the same cautions on the input and output capacitors, but an additional caution due to R_b.

R_b requires a bit of explanation. An ideal op amp would form a potential $+V_{REF}$ at its non-inverting input in response to the presence of $+V_{REF}$ at its inverting input. A real-world op amp, however, requires a small amount of input bias current at the non-inverting input. If R_b is not present, there is no source for input bias current.

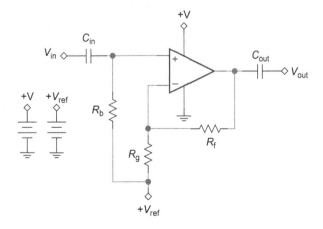

Figure 3.3
Single-Supply Non-Inverting Op Amp Gain Circuit

Therefore, the circuit would not work, at least not properly. The exact value of R_b is dependent on the input bias current specification in the data sheet, and some experimentation may be needed to get it right. It will produce a small DC offset superimposed on the output waveform, but that will be blocked by C_{OUT} along with $+V_{REF}$, which will appear at the output. In extreme cases, there may be enough DC offset generated by the input bias current to affect the voltage swing of the op amp. The easiest way to test for this is to look at the op amp output before C_{OUT}, referred to $+V_{REF}$. If it is skewed too greatly positive, or negative, try scaling R_b or select an op amp with lower input bias current.

3.3 Simultaneous Equations

When the DC must be amplified along with the AC, it introduces many more variables. There are many possible variations of inverting and non-inverting circuits that contain both DC and AC gain components. Rather than just hoping to stumble upon the one that solves the circuit problem, it is better to develop a systematic methodology that will yield an optimum solution. The simultaneous equation method can be used to render specified data into equation form. The goal is to use that equation to design an op amp circuit whose DC transfer (or DC sweep) matches the results obtained from the equation. An op amp DC transfer characteristic will form a straight line when plotted on Cartesian coordinates. For those not familiar with Cartesian coordinates or needing a refresher, look at Figure 3.4.

Cartesian coordinates are plotted on a two-dimensional graph, with an x-axis and a y-axis. By convention, the origin point (0,0) is plotted in the middle, the x-axis is horizontal, and the y-axis is vertical. The positive portion of the x-axis is on the right, and the positive portion of the y-axis is the upper portion of the graph. The location of any point on the graph can be represented by a pair of numbers (or coordinates), with x being first and y being second. The indicated point is 3 x units to the right of the origin, and 4 y units above the origin. The slope of a line drawn from the origin to the indicated point would be the gain in y divided by the gain in x, or 4/3.

We have actually already been using Cartesian coordinates in our discussion of inverting and non-inverting gains. The gains of an inverting op amp stage, for example, can be indicated by a diagonal line from quadrant 2 to quadrant 4 on the graph, going through the origin, with a slope equal to $-R_f/R_g$. Similarly, the gain of a non-inverting op amp stage would be indicated by a diagonal line from quadrant 3, through the origin, and into quadrant 1, with a slope equal to $1 + R_f/R_g$. I will talk more about these two circuits in a discussion later in this chapter. But for now, I will discuss cases where the line does not go through the origin, which means they have DC offset.

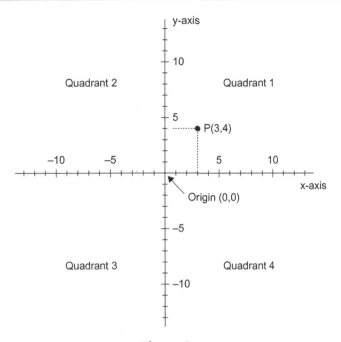

Figure 3.4
Cartesian Coordinates

I am going to represent the slope of a line on the Cartesian coordinates, and therefore the gain of the op amp, by the letter *m*, which can be either positive (non-inverting gain) or negative (inverting gain). The input voltage V_{IN} is represented by the x-axis, and the output voltage V_{OUT} is represented by the y-axis. Therefore, the transfer function of the op amp circuits discussed to date can be represented by Equation 3.1:

$$y = \pm mx \qquad (3.1)$$

where *m* for the inverting case is equal to $-R_f/R_g$, and *m* for the non-inverting case is equal to $1 + R_f/R_g$.

If the transfer equation is not constrained by the requirement to cross through the origin at (0,0), then the transfer equation contains an offset term I will call *b*, which is the point at which the resulting line will cross the y-axis. *b* corresponds to a DC offset that is superimposed on the gain function. Like the gain term *m*, it can be either positive or negative. Therefore, a linear op amp DC transfer function can be completely defined by the equation of a straight line (Equation 3.2):

$$y = \pm mx \pm b \qquad (3.2)$$

The equation of a straight line has four possible solutions depending on the sign of the slope m and the intercept b; thus simultaneous equations yield solutions in four forms. Four circuits must be developed, one for each form of the equation of a straight line. The four equations, cases, or forms of a straight line are given in Equations 3.3–3.6, where electronics terminology has been substituted for math terminology:

$$V_{OUT} = +mV_{IN} + b \qquad (3.3)$$

$$V_{OUT} = +mV_{IN} - b \qquad (3.4)$$

$$V_{OUT} = -mV_{IN} + b \qquad (3.5)$$

$$V_{OUT} = -mV_{IN} - b \qquad (3.6)$$

Given a set of two data points for V_{OUT} and V_{IN}, simultaneous equations are solved to determine m and b for the equation that satisfies the given data. The sign of m and b determines the type of circuit required to implement the solution. The given data is derived from the specifications; i.e. a sensor output signal ranging from 0.1 to 0.2 V must be interfaced into an analog-to-digital converter that has an input voltage range of 1–4 V. These data points ($V_{OUT} = 1$ V at $V_{IN} = 0.1$ V, $V_{OUT} = 4$ V at $V_{IN} = 0.2$ V) are inserted into Equation 3.3, as shown in Equations 3.7 and 3.8, to obtain m and b for the specifications:

$$1 = m(0.1) + b \qquad (3.7)$$

$$4 = m(0.2) + b \qquad (3.8)$$

Multiply Equation 3.7 by 2 and subtract it from Equation 3.8:

$$2 = m(0.2) + 2b \qquad (3.9)$$

$$b = -2 \qquad (3.10)$$

After algebraic manipulation of Equation 3.7, substitute Equation 3.10 into Equation 3.7 to obtain Equation 3.11:

$$m = \frac{2+1}{0.1} = 30 \qquad (3.11)$$

Now m and b are substituted back into Equation 3.3, yielding Equation 3.12:

$$V_{OUT} = 30V_{IN} - 2 \qquad (3.12)$$

Notice that, although Equation 3.3 was the starting point, the form of Equation 3.12 is identical to the format of Equation 3.4. The specifications or given data determine the sign of m and b, and starting with Equation 3.3, the final equation form is discovered after m and b are calculated. The next step required to complete the problem solution is to develop a circuit that has $m = 30$ and $b = -2$. Circuits were developed for Equations 3.3–3.6, and they are given the names Case 1–4, respectively. There are different circuits that will yield the same equations, but these circuits were selected because they do not require negative voltage references which are hard to obtain.

Note that I do not show decoupling capacitors in any of the examples to follow. You are responsible for adding appropriate decoupling to each circuit.

I will start out doing very detailed math, to familiarize you with the theoretical process. As you read on in the chapter, I will scale down the math to more manageable levels. Please bear with me at first!

3.3.1 Case 1: $V_{OUT} = +mV_{IN} + b$

The circuit configuration that yields a solution for Case 1 is shown in Figure 3.5. Notice right away that it is sort of intuitive that when m and b are both positive, they would be applied to the non-inverting input of the op amp. This will also be obvious in subsequent cases. If the op amp is selected properly, this circuit can be operated from a single supply.

The circuit equation is written using the voltage divider rule and superposition:

$$V_{OUT} = V_{IN}\left(\frac{R_2}{R_1 + R_2}\right)\left(\frac{R_f + R_g}{R_g}\right) + V_{REF}\left(\frac{R_1}{R_1 + R_2}\right)\left(\frac{R_f + R_g}{R_g}\right) \qquad (3.13)$$

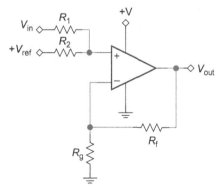

Figure 3.5
Schematic for Case 1: $V_{OUT} = +mV_{IN} + b$

The equation of a straight line (Case 1) is repeated in Equation 3.14 so that comparisons can be made between it and Equation 3.13:

$$V_{OUT} = mV_{IN} + b \tag{3.14}$$

Equating coefficients yields Equations 3.15 and 3.16:

$$m = \left(\frac{R_2}{R_1 + R_2}\right)\left(\frac{R_f + R_g}{R_g}\right) \tag{3.15}$$

$$b = V_{REF}\left(\frac{R_1}{R_1 + R_2}\right)\left(\frac{R_f + R_g}{R_g}\right) \tag{3.16}$$

Case 1 Example

Consider a circuit where the design requirements are $V_{OUT} = 1$ V at $V_{IN} = 0.01$ V, $V_{OUT} = 4.5$ V at $V_{IN} = 1$ V, $R_L = 10$ k, and $V_{CC} = 5$ V. Assume that V_{CC} is also used for the reference input, and therefore $V_{REF} = 5$ V. This is probably not the best idea, as it sacrifices noise performance, accuracy, and stability performance.

Nevertheless, this is often done when cost is an important specification, but the V_{CC} supply must be specified well enough to do the job. Each step in the subsequent design procedure is included in this analysis to ease learning, although it may get a bit tedious. Therefore, intermediate steps are skipped when subsequent cases are analyzed.

The data is substituted into simultaneous equations:

$$1 = m(0.01) + b \tag{3.17}$$

$$4.5 = m(1.0) + b \tag{3.18}$$

Equation 3.17 is multiplied by 100 (Equation 3.19) and Equation 3.18 is subtracted from Equation 3.19 to obtain Equation 3.20:

$$100 = m(1.0) + 100b \tag{3.19}$$

$$b = \frac{95.5}{99} = 0.9646 \tag{3.20}$$

The slope of the transfer function, m, is obtained by substituting b into Equation 3.17:

$$m = \frac{1 - b}{0.01} = \frac{1 - 0.9646}{0.01} = 3.535 \tag{3.21}$$

Now that b and m have been calculated, the resistor values can be calculated. Equations 3.15 and 3.16 are solved for the quantity $(R_f + R_g)/R_g$, and then they are set equal in Equation 3.22, thus yielding Equation 3.23:

$$\frac{R_f + R_g}{R_g} = m\left(\frac{R_1 + R_2}{R_2}\right) = \frac{b}{V_{CC}}\left(\frac{R_1 + R_2}{R_1}\right) \tag{3.22}$$

$$R_2 = \frac{3.535}{0.9646/5}R_1 = 18.316R_1 \tag{3.23}$$

Although it is seldom a good idea, 5% tolerance resistors are used in this example, so we choose $R_1 = 10$ kΩ, and that sets the value of $R_2 = 183.16$ kΩ. The closest 5% resistor value to 183.16 kΩ is 180 kΩ; therefore, select $R_1 = 10$ kΩ and $R_2 = 180$ kΩ. Being forced to choose standard resistor values means that there is an error in the circuit transfer function, because m and b are not exactly the same as calculated. The real world constantly forces compromises into circuit design. Resistor values closer to the calculated values could be selected by using 1% resistors. The left half of Equation 3.22 is used to calculate R_f and R_g:

$$\frac{R_f + R_g}{R_g} = m\left(\frac{R_1 + R_2}{R_2}\right) = 3.535\left(\frac{180 + 10}{180}\right) = 3.73 \tag{3.24}$$

$$R_f = 2.73R_g \tag{3.25}$$

The resulting circuit equation is given below:

$$V_{OUT} = 3.5V_{IN} + 0.97 \tag{3.26}$$

The gain setting resistor, R_g, is selected as 10 kΩ, and 27 kΩ, the closest 5% standard value is selected for the feedback resistor, R_f. Again, there is an error involved with standard resistor values. The circuit with the selected component values and transfer curve is shown in Figure 3.6.

The transfer curve in Figure 3.6 shows a portion of quadrant 1 of Cartesian coordinates. In this case, there can be no line in the other quadrants, because that would require a negative power supply. The transfer curve shown is a straight line, and that means that the circuit is linear. The V_{OUT} intercept is about 0.98 V rather than 1 V as specified. The output voltage measured 4.53 V when the input voltage was 1 V. This is excellent performance considering that the resistor sequence selected was 5%. But do not be fooled into complacency. These results are nominal results if the resistor values were perfect. Different sets of components will have slightly different intercepts and slopes because of the resistor tolerances.

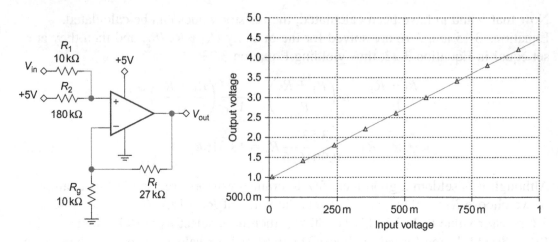

Figure 3.6
Case 1 Example Circuit

Output Voltage Swing

This is an excellent time to introduce a real-world op amp limitation: the output voltage swing specifications. Obviously, the op amp output cannot swing beyond the power supply rails. It also cannot reach them. The output transistors inside the op amp are real-world components as well, and their voltage drop on the high and low ends constrains the output voltage to a level less than the op amp's most positive rail, and above the op amp's negative rail. These parameters are listed on the data sheet as V_{OH} and V_{OL}, respectively. They may or may not be the same value, so both high- and low-voltage excursions must be analyzed. Older generations of op amps have large values of V_{OH} and V_{OL}, with 3 V or more in some cases (such as high-voltage op amps). Obviously, values of 3 V are acceptable when an op amp is operated off ± 15 V, and would limit the output voltage swing to ± 12 V. The same op amp operated off ± 5 V supplies would be limited to an output voltage swing of ± 2 V, a severe constraint that might be acceptable for small signal levels; but an op amp with 3 V values of V_{OH} and V_{OL} would be unsuitable for operation off a single 5 V supply, because the values of V_{OH} and V_{OL} will not allow the output to work at all! As power supply voltages have decreased in portable devices, a new class of "rail-to-rail" op amps has been developed by semiconductor manufacturers. You should be wary of the term and check the data sheet for the real values of V_{OH} and V_{OL}. There is no such thing as a true "rail-to-rail output" op amp. It can be close, but like the case with open-loop gain, the output voltage swing can never reach the rails, because the output transistors inside the op amp require some voltage for internal bias.

Back to the Case 1 Example

This circuit must have an output voltage swing from 1 V to 4.5 V. A TLV2471 is selected. The table for operation from +5 V is consulted on the data sheet. The V_{OH} and V_{OL} values are affected not only by temperature, but also by output load. You must take into account the load the op amp is driving, as well as the operating temperature range. For this example, the lowest V_{OH} specification is 4.65 V, which means that the circuit will be able to swing to the required 4.5 V under all data sheet conditions. Similarly, the highest V_{OL} specification is 0.35 V, which means that the circuit will be able to swing to the required 1 V under all data sheet conditions.

3.3.2 Case 2: $V_{OUT} = +mV_{IN} - b$

The circuit shown in Figure 3.7 yields a solution for Case 2. The circuit equation is obtained by taking the Thevenin equivalent circuit looking into the junction of R_1 and R_2. After the R_1, R_2 circuit is replaced with the Thevenin equivalent circuit, the gain is calculated with the ideal gain equation (Equation 3.27):

$$V_{OUT} = V_{IN}\left(\frac{R_f + R_g + R_1||R_2}{R_g + R_1||R_2}\right) - V_{REF}\left(\frac{R_2}{R_1 + R_2}\right)\left(\frac{R_f}{R_g + R_1||R_2}\right) \quad (3.27)$$

Comparing terms in Equations 3.27 and 3.4 enables the extraction of *m* and *b*:

$$m = \frac{R_f + R_g + R_1||R_2}{R_g + R_1||R_2} \quad (3.28)$$

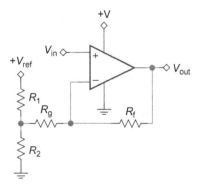

Figure 3.7
Schematic for Case 2: $V_{OUT} = +mV_{IN} - b$

$$|b| = V_{REF}\left(\frac{R_2}{R_1 + R_2}\right)\left(\frac{R_f}{R_g + R_1||R_2}\right) \tag{3.29}$$

Case 2 Example

Consider a circuit where the design requirements are: $V_{OUT} = 1.5$ V at $V_{IN} = 0.2$ V, $V_{OUT} = 4.5$ V at $V_{IN} = 0.5$ V, $V_{REF} = V_{CC} = 5$ V, $R_L = 10$ kΩ, and 5% resistor tolerances. The simultaneous equations (Equations 3.30 and 3.31) are:

$$1.5 = 0.2m + b \tag{3.30}$$

$$4.5 = 0.5m + b \tag{3.31}$$

From these equations we find that $b = -0.5$ and $m = 10$. Making the assumption that $R_1||R_2 << R_g$ simplifies the calculations of the resistor values.

$$m = 10 = \frac{R_f + R_g}{R_g} \tag{3.32}$$

$$R_f = 9R_g \tag{3.33}$$

Let $R_g = 20$ kΩ, and then $R_f = 180$ kΩ.

$$b = V_{CC}\left(\frac{R_f}{R_g}\right)\left(\frac{R_2}{R_1 + R_2}\right) = 5\left(\frac{180}{20}\right)\left(\frac{R_2}{R_1 + R_2}\right) \tag{3.34}$$

$$R_1 = \frac{1 - 0.01111}{0.01111}R_2 = 89R_2 \tag{3.35}$$

Select $R_2 = 820\Omega$ and R_1 equals 72.98 kΩ. Since 72.98 kΩ is not a standard 5% resistor value, R_1 is selected as 75 kΩ. The difference between the selected and calculated value of R_1 has about a 3% effect on b, and this error shows up in the transfer function as an intercept rather than a slope error. The parallel resistance of R_1 and R_2 is approximately 820Ω and this is much less than R_g, which is 20 kΩ, thus the earlier assumption that $R_g >> R_1||R_2$ is justified. R_2 could have been selected as a smaller value, but the smaller values yielded poor standard 5% values for R_1. The final circuit and transfer curves are shown in Figure 3.8.

The TLV2471 can also be used to build the test circuit because of its wide output voltage swing.

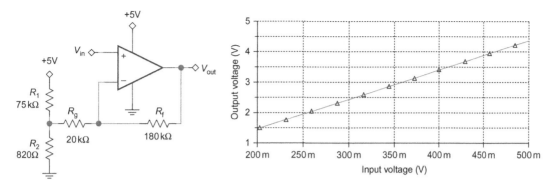

Figure 3.8
Case 2 Example Circuit

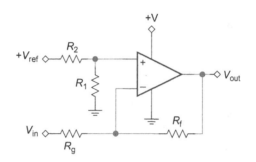

Figure 3.9
Schematic for Case 3: $V_{OUT} = +mV_{IN} + b$

3.3.3 Case 3: $V_{OUT} = -mV_{IN} + b$

The circuit shown in Figure 3.9 yields the transfer function desired for Case 3. The circuit equation is obtained with superposition.

$$V_{OUT} = -V_{IN}\left(\frac{R_f}{R_g}\right) + V_{REF}\left(\frac{R_1}{R_1 + R_2}\right)\left(\frac{R_f + R_g}{R_g}\right) \tag{3.36}$$

Comparing terms between Equations 3.36 and 3.5 enables the extraction of m and b.

$$|m| = \frac{R_f}{R_g} \tag{3.37}$$

$$b = V_{REF}\left(\frac{R_1}{R_1 + R_2}\right)\left(\frac{R_f + R_g}{R_g}\right) \tag{3.38}$$

Case 3 Example

Consider a circuit where the design requirements are: $V_{OUT} = 1$ V at $V_{IN} = -0.1$ V, $V_{OUT} = 6$ V at $V_{IN} = -1$ V, $V_{REF} = V_{CC} = 10$ V, $R_L = 100\Omega$, and 5% resistor tolerances. The simultaneous equations (Equations 3.39 and 3.40) are:

$$1 = (-0.1)m + b \tag{3.39}$$

$$6 = (-1)m + b \tag{3.40}$$

From these equations we find that $b = 0.444$ and $m = -5.6$.

$$|m| = 5.56 = \frac{R_f}{R_g} \tag{3.41}$$

$$R_f = 5.56 R_g \tag{3.42}$$

Let $R_g = 10$ kΩ, and then $R_f = 56$ kΩ.

$$b = V_{CC}\left(\frac{R_f + R_g}{R_g}\right)\left(\frac{R_1}{R_1 + R_2}\right) = 10\left(\frac{56 + 10}{10}\right)\left(\frac{R_1}{R_1 + R_2}\right) \tag{3.43}$$

$$R_2 = \frac{66 - 0.4444}{0.4444}R_1 = 147.64R_1 \tag{3.44}$$

The final equation for the example is:

$$V_{OUT} = -5.56V_{IN} + 0.444 \tag{3.45}$$

Select $R_1 = 2$ kΩ and $R_2 = 300$ kΩ. The difference between the selected and calculated value of R_1 has a nearly insignificant effect on b. The final circuit and transfer curve are shown in Figure 3.10.

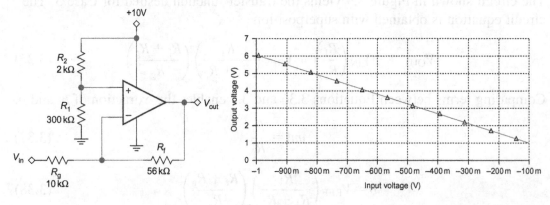

Figure 3.10
Case 3 Example Circuit

There is an issue when V_{CC} is powered down while there is a negative voltage on the input circuit; most of the negative voltage appears on the inverting op amp input lead. The easiest solution is to connect a diode, with its cathode on the inverting op amp input lead and its anode at ground. If a negative voltage gets on the inverting op amp input lead, it is clamped to ground by the diode. Select the diode type as germanium or Schottky so the voltage drop across the diode is about 200 mV; this small voltage does not harm most op amp inputs. As a further precaution, R_g can be split into two resistors with the diode inserted at the junction of the two resistors. This places a current limiting resistor between the diode and the inverting op amp input lead.

3.3.4 Case 4: $V_{OUT} = -mV_{IN} - b$

The circuit shown in Figure 3.11 yields a solution for Case 4. The circuit equation is obtained by using superposition to calculate the response to each input. The individual responses to V_{IN} and V_{REF} are added to obtain Equation 3.46:

$$V_{OUT} = -V_{IN}\frac{R_f}{R_{g1}} - V_{REF}\frac{R_f}{R_{g2}}$$ (3.46)

Comparing terms in Equations 3.56 and 3.16 enables the extraction of m and b.

$$|m| = \frac{R_f}{R_{g1}}$$ (3.47)

$$|b| = V_{REF}\frac{R_f}{R_{g2}}$$ (3.48)

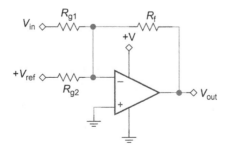

Figure 3.11
Schematic for Case 4: $V_{OUT} = -mV_{IN} - b$

Case 4 Example

Consider a circuit where the design requirements are: $V_{OUT} = 1$ V at $V_{IN} = -0.1$ V, $V_{OUT} = 5$ V at $V_{IN} = -0.3$ V, $V_{REF} = V_{CC} = 5$ V, $R_L = 10$ kΩ, and 5% resistor tolerances. The simultaneous equations (Equations 3.49 and 3.50) are:

$$1 = (-0.1)m + b \qquad (3.49)$$

$$5 = (-0.3)m + b \qquad (3.50)$$

From these equations we find that $b = -1$ and $m = -20$. Setting the magnitude of m equal to Equation 3.47 yields Equation 3.51:

$$|m| = 20 = \frac{R_f}{R_{g1}} \qquad (3.51)$$

$$R_f = 20R_{g1} \qquad (3.52)$$

Let $R_{g1} = 1$ kΩ, and then $R_f = 20$ kΩ.

$$|b| = V_{CC}\left(\frac{R_f}{R_{g1}}\right) = 5\left(\frac{R_f}{R_{g2}}\right) = 1 \qquad (3.53)$$

$$R_{g2} = \frac{R_f}{0.2} = \frac{20}{0.2} = 100 \text{ k}\Omega \qquad (3.54)$$

The final equation for this example is given in Equation 3.55:

$$V_{OUT} = -20V_{IN} - 1 \qquad (3.55)$$

The final circuit and transfer curve are shown in Figure 3.12.

As before, there may be an issue with power sequencing, the input voltage being present before power is applied to the op amp. This can be addressed again with a diode on the inverting input of the op amp.

3.4 So, Where to Now?

These original four cases were presented by Ron Mancini in the book *Op Amps for Everyone*, a book on which I (Bruce Carter) collaborated. Further work, and John Bishop and I, established an entire continuum of circuits which can handle attenuation as well as amplification, and incorporate already familiar circuits and applications. Some of these required new topologies entirely. When tabularized, the complete set of cases looks like the layout shown in Table 3.1.

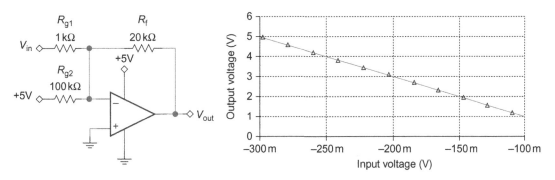

Figure 3.12
Case 4 Example Circuit

Table 3.1: Comprehensive List of Cases

		Offset <0	Offset $=0$	Offset >0
	Gain >1	Case $=2$	Case $=5$	Case $=1$
Non-inverting	Gain $=1$	Case $=8$	Case $=9$	
	$0<$ Gain <1		Case $=10$	Case $=7$
	Gain $=0$	Regulators (see Chapter 10)		
	$-1<$ Gain <0	Case $=12$	Case $=13$	Case $=11$
Inverting	Gain ≤ -1	Case $=4$	Case $=6$	Case $=3$

The intention of Table 3.1 is to define every possible combination of gain and offset possible for an op amp circuit. Some of these cases have already been discussed. Ron's four cases appear in the corners of the final three columns in Table 3.1. Cases 5 and 6 have been covered in Chapter 2. Case 9 is shown in Figure 3.13 and is a familiar case to most designers: the non-inverting buffer. Case 9 is easy to understand. m (the gain) equals 1 and b (the offset) equals zero. It is included as its own case because it is separate and distinct from the non-inverting gain stage analyzed in Chapter 2.

A slight variation of Case 9 occurs when you place a voltage divider on V_{IN} as shown in Figure 3.14. This creates the first of the attenuation cases, Case 10. It has no offset, so b equals zero. m is determined by the voltage divider equation:

$$V_{OUT} = m \times V_{IN}$$

$$m = \frac{R_2}{R_1 + R_2} \tag{3.56}$$

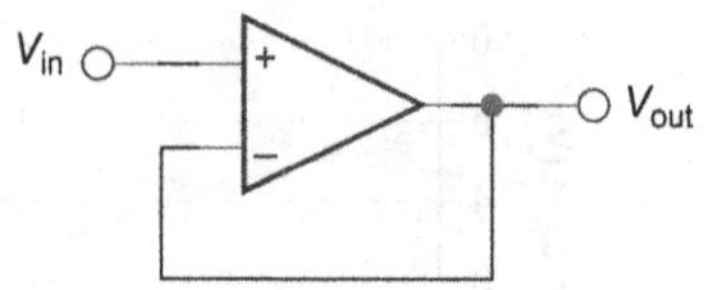

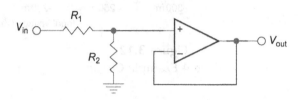

Figure 3.13
Case 9 Circuit

Figure 3.14
Case 10 Circuit

These are the simplest of the additional cases. Cases 7, 8, 11, 12, and 13 are more esoteric and should be encountered less frequently. They exist because of the stability criteria of op amps. If an op amp is described in the data sheet as "unity gain stable", you should realize that unity gain is the least stable operating point. The stage is actually more stable above unity gain. The late William Ezell came up with a scheme for handling inverting attenuation op amp circuits based on a "T" attenuation pad, shown in Figure 3.15. This forms the core of Cases 11, 12, and 13, with Case 13 (no offset) being the simplest.

Case 11 (Figure 3.16) adds positive offset into the non-inverting input, and Case 12 (Figure 3.17) adds negative offset to the inverting input by summing with the signal. This leaves Cases 7 and 8, non-inverting attenuations with offset. It would be nice if they could be slight variations on the very simple Case 10, but that is not so. Case 8 (Figure 3.18) bears a superficial resemblance to Case 3, except that the locations of V_{REF} and V_{IN} have been swapped. Case 7 (Figure 3.19), however, proves to be deceptively simple, yet the algorithm to calculate it maddeningly difficult.

3.5 A Design Procedure, and a Design Aid

The design procedure for single-supply op amp design is:

1. Substitute the design requirements into simultaneous equations to obtain m and b (the slope and intercept of a straight line).
2. Let m and b determine the form of the circuit.

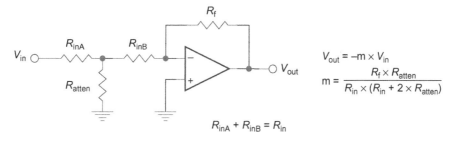

Figure 3.15
Case 13 (Inverting Attenuator) Circuit

$$V_{out} = -m \times V_{in}$$

$$m = \frac{R_f \times R_{atten}}{R_{in} \times (R_{in} + 2 \times R_{atten})}$$

$$R_{inA} + R_{inB} = R_{in}$$

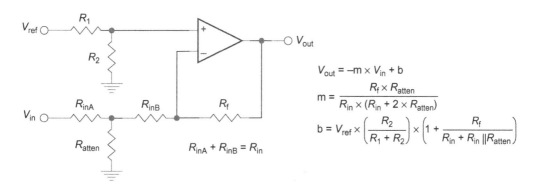

Figure 3.16
Case 11 Circuit

$$V_{out} = -m \times V_{in} + b$$

$$m = \frac{R_f \times R_{atten}}{R_{in} \times (R_{in} + 2 \times R_{atten})}$$

$$b = V_{ref} \times \left(\frac{R_2}{R_1 + R_2}\right) \times \left(1 + \frac{R_f}{R_{in} + R_{in} \| R_{atten}}\right)$$

$$R_{inA} + R_{inB} = R_{in}$$

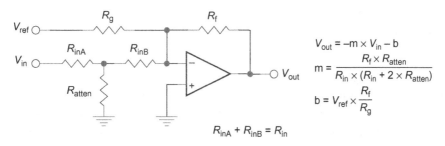

Figure 3.17
Case 12 Circuit

$$V_{out} = -m \times V_{in} - b$$

$$m = \frac{R_f \times R_{atten}}{R_{in} \times (R_{in} + 2 \times R_{atten})}$$

$$b = V_{ref} \times \frac{R_f}{R_g}$$

$$R_{inA} + R_{inB} = R_{in}$$

3. Choose the circuit configuration that fits the form.
4. Using the circuit equations for the circuit configuration selected, calculate the resistor values.
5. Build the circuit, take data, and verify performance.

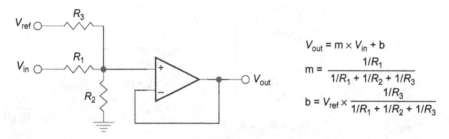

Figure 3.18
Case 8 Circuit

$$V_{out} = m \times V_{in} - b$$

$$m = \left(\frac{R_2}{R_1 + R_2}\right) \times \left(1 + \frac{R_f}{R_g}\right)$$

$$b = V_{ref} \times \frac{R_f}{R_g}$$

Figure 3.19
Case 7 Circuit

$$V_{out} = m \times V_{in} + b$$

$$m = \frac{1/R_1}{1/R_1 + 1/R_2 + 1/R_3}$$

$$b = V_{ref} \times \frac{1/R_3}{1/R_1 + 1/R_2 + 1/R_3}$$

6. Test the circuit for nonstandard operating conditions (circuit power off while interface power is on, over/under range inputs, etc.).
7. Add protection components as required.
8. Retest.

The biggest problem you face is the first and second steps above. Fortunately, if you know the circuit power supply voltage(s), the reference voltage, the input voltage high and low values, and the output voltage low and high values, the case is already defined, but the definition is not apparent. I have given you 13 cases, but you have no easy procedure to find out which case you are dealing with. Rather than leave you hanging, I will provide the single schematic of Figure 3.20, and a design aid that will automatically determine the case.

Although not obvious yet, this single circuit can implement all 13 cases above. When used to implement a particular case, some resistors will be left open, some will be shorted through 0Ω jumper resistors, and some will be values associated with the case. I have written a Javascript calculator which can be accessed on the companion website for this volume. A screen shot is shown in Figure 3.21.

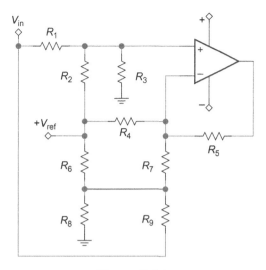

Figure 3.20
Universal Op Amp Circuit

Given the inputs for V_{IN} low and high, V_{OUT} low and high, and the reference voltage, the calculator will automatically determine the case, gain (m), offset (b), and the value for all nine resistors in the schematic above. It will also give the real outputs of gain, offset, V_{OUT} zero scale, and V_{OUT} full scale that can be expected with the resistor values it calculates. It comes preloaded with the test cases above. These preloaded values can be overwritten. If you want to start with a test case to get a feel for the calculator, simply use the drop-down menu next to the test case button, and click the test case button. This loads the top fields. Then click the calculate button. When you are ready for your design, overwrite the top fields and hit calculate. The calculator automatically updates the line equation and expressions for m and b for each case. It is that easy!

A couple of notes:

- Case 7, as I stated above, was deceptively difficult. It requires a seed value, which you can select in a pull-down menu of 1% sequence resistors. A solution to Case 7 will appear, but you might want to try different values of seed resistor to optimize m, b, or both.
- The order of magnitude of the resistor set can be changed with a multiplier drop-down selection in the top fields. In most cases it can be left alone, but if you do not like the order of magnitude of the resistors, you can recalculate quickly after selecting a multiplier by simply clicking the calculate box again.

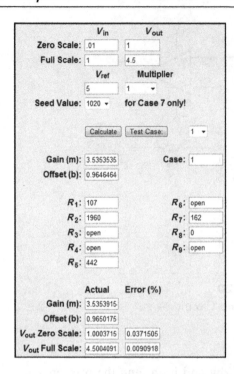

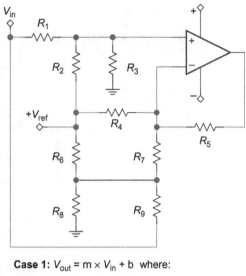

Case 1: $V_{out} = m \times V_{in} + b$ where:

$$m = \left(\frac{R_2}{R_1 + R_2}\right) \times \left(1 + \frac{R_5}{R_7}\right)$$

$$b = V_{ref} \times \left(\frac{R_1}{R_1 + R_2}\right) \times \left(1 + \frac{R_5}{R_7}\right)$$

Figure 3.21
Universal Op Amp Calculator

The resistors below are automatically scaled to the new multiplier. A simple function, but one I find handy.

To further facilitate your designs, I have done a board layout (Figure 3.22) of the universal schematic above. It is designed to accommodate SO-8 single op amps and 1206 surface-mount resistors. It is implemented on a single printed circuit board layer. The gerbers are available from the companion website.

This is a universal topology that will allow any transducer to interface with any data converter (within reason, of course). If you have no idea what the final interface should be, you can implement this circuit topology very early in the design cycle, knowing that when final decisions are eventually made, this circuit has a very good chance of being adaptable to the situation.

3.6 Summary

Single-supply op amp design is more complicated than split-supply op amp design, but with a logical design approach excellent results are achievable. Single-supply

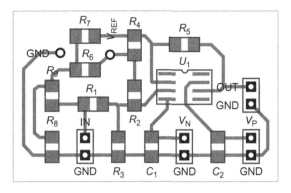

Figure 3.22
Universal Op Amp Board

design used to be considered technically limiting because older op amps had limited capability. New op amps have excellent single-supply parameters; thus, when used in the correct applications these op amps yield rail-to-rail performance equal to that of their split-supply counterparts.

Single-supply op amp design usually involves some form of biasing, and this requires more thought, so single-supply op amp design needs discipline and a procedure. When this procedure is followed, good results follow. Remember, the only equation a linear op amp can produce is the equation of a straight line. Every case above produces a result that is a straight line.

Figure 2.22
Small Op-Amp Box...

designed to be considered separately handle... because under op-amps had limited capability. New op-amps have solved our single-supply parameter... that, when used in the correct application, these op-amps yield excellent performance equal to that of their split-supply counterparts.

Single-supply op-amp design usually involves some form of biasing, and this requires more thought, so single-supply op-amp "might need a discipline and a procedure. When this procedure is followed, a good result... follow. Remember, the op-amp equation for a linear op-amp can produce the equation of a straight line. Every case above produces a result that is a straight line.

CHAPTER 4

Different Types of Op Amps

4.1 Voltage Feedback Op Amps

The op amps discussed in the previous chapters are voltage feedback devices. They represent the original type of op amp, which is overwhelmingly the most common type in use. You may very well go an entire career and never encounter a different type of op amp. In a voltage feedback op amp, a portion of the voltage output is fed back into the input out of phase with the other input by 180°, resulting in cancellation of the majority of the gain. This also makes the gain dependent primarily on two relatively stable resistors instead of the active components in the op amp.

With few exceptions, the previous chapters assumed an ideal op amp model. The name *ideal op amp* is applied because the salient parameters of the op amp are assumed to be perfect. There is no such thing as an ideal op amp, but present day op amps come so close to ideal that ideal op amp analysis approaches actual analysis. In addition, when working at low frequencies, several kilohertz, the ideal op amp analysis produces accurate answers.

Table 4.1 lists the basic ideal op amp assumptions and Figure 4.1 shows where they occur in the ideal op amp.

For more information on these parameters and what they mean, consult Appendix A. For now, they are introduced so the reader can understand the differences between a voltage feedback amplifier (VFA) and the current feedback amplifier (CFA) of Section 4.3.

4.2 Uncompensated/Undercompensated Voltage Feedback Op Amps

I include these as a separate type of op amp because they add a level of complexity to the task of design, and also because they introduce an important topic: stability.

The VFAs discussed up to this point were assumed to be "unity gain stable". So, what does that term mean? Figure 4.2 (the same as Figure 2.10 in Chapter 2) shows a compensated op amp, which for purposes of this discussion I will say is unconditionally stable at unity gain (0 dB), because its internal construction consists

Table 4.1: Basic Ideal Op Amp Assumptions

Parameter Name	Parameter Symbol	Value
Input current	I_{IN}	0
Input offset voltage	V_{OS}	0
Input impedance	Z_{IN}	∞
Output impedance	Z_{OUT}	0
Gain	a	∞

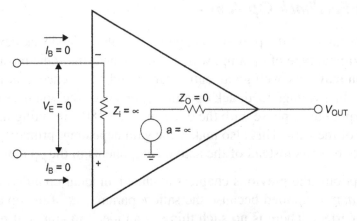

Figure 4.1
The Ideal Op Amp

of a single low-pass pole that keeps the slope of the response to -20 dB per decade. This is not always the case, as Figure 4.3 (adapted from a real data sheet) shows.

In this case, the slope of the gain curve takes an additional dive below 20 dB, to a slope of -40 dB per decade or perhaps more. Why does this happen? In an effort to make op amps "faster", some manufacturers change the internal compensation network to introduce less roll-off. This increases the open-loop gain region of the op amp to include higher frequencies, but the trade-off is that the op amp is no longer stable at unity gain. In this case, the op amp would be stable at a gain of 10 (20 dB), but no lower. If used at unity gain, it would be in danger of sustained oscillation, which requires a brief explanation from loop theory.

The basic loop equation, as applied to electronics, is shown in Figure 4.4. The output equation is written in Equation 4.1.

$$V_{OUT} = EA \tag{4.1}$$

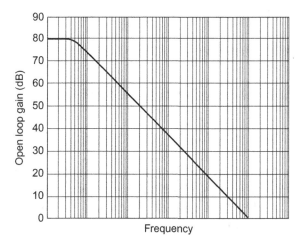

Figure 4.2
Op Amp Open-Loop Response

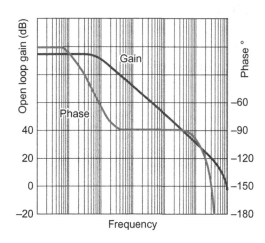

Figure 4.3
Undercompensated Op Amp

The error equation is written in Equation 4.2:

$$E = V_{IN} - \beta V_{OUT} \tag{4.2}$$

Combining Equations 4.1 and 4.2 yields Equation 4.3:

$$\frac{V_{OUT}}{A} = V_{IN} - \beta V_{OUT} \tag{4.3}$$

Collecting terms yields Equation 4.4:

$$V_{OUT}\left(\frac{1}{A} + \beta\right) = V_{IN} \tag{4.4}$$

Rearranging terms yields the classic form of the feedback equation (Eqn 4.5):

$$\frac{V_{OUT}}{V_{IN}} = \frac{A}{1 + A\beta} \tag{4.5}$$

There is a potential instability when the term $A\beta$, also called the loop gain, becomes equal to -1. Now, keep in mind that we are using complex numbers, which have magnitude and direction. The loop gain can approach -1, when it is mathematically the complex number $1\angle - 180°$. Equation 4.5 then approaches $1/0 \Rightarrow \infty$. The circuit output heads for infinity as fast as it can using the equation of a straight line. If the output were not energy limited, the circuit would explode the world, but happily, it is energy limited, so somewhere it comes up against the limit.

Referring to Figure 4.5, which is the loop analysis diagram from Figure 4.4 modified to show a non-inverting op amp stage, you should realize that the Σ and A blocks are internal to the op amp, while the β block is the feedback network consisting of two resistors R_f and R_g forming a voltage divider to the " $-$ " input of the Σ block (which is the input stage of the op amp).

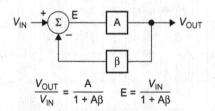

Figure 4.4
Electronics Version of the Feedback Diagram and Equations

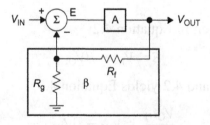

Figure 4.5
The Feedback Loop Analysis Figure Modified to Show a Non-Inverting Op Amp Stage

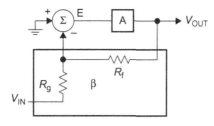

Figure 4.6
The Feedback Loop Analysis Figure Modified to Show an Inverting Op Amp Stage

The analysis of the familiar inverting op amp gain stage is a bit more complicated, but you can think of the bottom of the voltage divider resistor R_g going to the input signal instead of to the " + " input of the Σ block (Figure 4.6).

So, the non-inverting and inverting stages have just been shown in a different way; this was done to show how they relate back to the loop gain equation.

You might be asking at this point: which components can cause the loop gain to have a $-180°$ phase shift? Figures 4.5 and 4.6 only show resistors, so there would be no phase shift. The phase shift comes from parasitic capacitors that are internal to the op amp. The effect of these internal parasitic capacitors accumulates the higher in frequency the amplifier goes, and results in an open-loop response similar to that in Figure 4.3. You could simply make sure that your circuit operates at a gain that would place the op amp in a stable operating zone. However, the history of op amp designs has proved that designers are notoriously ignorant about the topic of stability. Semiconductor companies are continually blamed for misapplication of op amps. For that reason, there are almost no uncompensated op amps sold at present, and very few undercompensated op amps. Semiconductor companies purposely introduce a low-pass pole inside the op amp which will swamp the effect of internal parasitic capacitance by rolling off the bandwidth before the internal parasitic capacitors can take effect, resulting in op amp responses similar to those shown in Figure 4.2. There is a single -20 dB per decade slope to the open-loop characteristic down to the rated gain of the op amp. Usually it is unity gain, but in a few cases, as in that of Figure 4.3, it is higher. **Key point: You are cautioned never to operate an op amp at a gain less than 1, or a gain less than the rated gain of an undercompensated amplifier.** This will be covered in more detail in Chapter 13: Common Application Mistakes.

Fortunately, there is an easy way for you to tell how close you are coming to instability at any gain. Referring to Figure 4.3, which shows the open-loop response of an undercompensated op amp, the phase response is also shown. **Key point: From what we just learned from loop theory, we want to avoid a phase shift of −180°.** The difference between −180° and the phase at the point where the open-loop gain crosses the rated gain of the op amp (in this case 20 dB) is a term called the phase margin. In this case it would be about 60°, which is a fairly safe value. The smaller the phase margin gets, the less stable the op amp stage will be. If the op amp of Figure 4.3 were to be operated at a gain of 40 dB, the phase margin would be 90°, and the stage would be even more stable. If the op amp of Figure 4.3 were to be operated at a gain of less than 20 dB, the phase margin would quickly drop to 0° and oscillation (or at least latch up at one or the other supply rails).

4.3 Current Feedback Op Amps

The CFA model is shown in Figure 4.7. The non-inverting input of a CFA connects to the input of the input buffer, so it has very high impedance − similar to that of a non-inverting VFA input. The inverting input connects to the input buffer's output, so the inverting input impedance is equivalent to a buffer's output impedance, which is very low. Z_B models the input buffer's output impedance, and it is usually less than 50Ω. The input buffer gain, G_B, is as close to one as integrated circuit (IC) design methods can achieve, and it is small enough to neglect in the calculations. The vastly different input impedances of the two inputs combined with DC offset errors inherent in the G_B buffer make CFAs unsuitable for differential and precision DC. The input buffer's output impedance cannot be ignored because it affects stability at high frequencies.

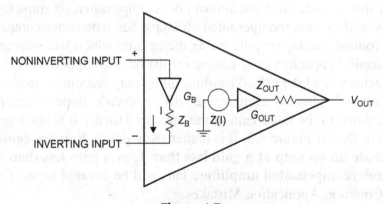

Figure 4.7
Current Feedback Amplifier Model

The output buffer provides low output impedance for the amplifier. Again, the output buffer gain, G_{OUT}, is very close to one, so it is neglected in the analysis. The output impedance of the output buffer is ignored during the calculations. The current-controlled current source, Z, is a transimpedance. The transimpedance in a CFA serves the same function as gain in a VFA; it is the parameter that makes the performance of the op amp dependent only on the passive parameter values. Usually the transimpedance is very high, in the megaohm range, so the CFA gains accuracy by closing a feedback loop in the same manner that the VFA does. In practice, the familiar circuit topologies used with VFAs are just as applicable for CFAs. There is one important distinction, however. While the absolute values of R_f and R_g have little effect on the stability of a VFA over several decades of values, the stability of a CFA is strongly dependent on the selection of R_f. The value for R_f at several gains is suggested in the data sheet: depart from the recommendation at the peril of an unstable circuit. This also means that you should never make a non-inverting buffer out of a CFA by merely connecting the output to the inverting input with a short. Create a non-inverting buffer by using the recommended value of R_f for unity gain from the data sheet.

There is one other important rule to remember with a CFA. Never connect a capacitor directly from output to inverting input. This rules out a number of active filter topologies (see Chapter 6). A resistor in series with a small capacitor is permissible, but exercise caution!

4.4 Fully Differential Op Amps

Op amps started out as fully differential components over 50 years ago. Techniques about how to use the fully differential versions have been almost lost over the decades in favor of single-ended op amps. Today's fully differential op amps offer performance advantages unheard of in those first units.

This chapter will just present the facts you need to get started with fully differential op amps, and some resources for further design assistance.

4.4.1 What Does "Fully Differential" Mean?

You should already be familiar with single ended op amps after reading the other chapters of this book. In brief, single-ended op amps have two inputs — a positive and a negative input — which are understood to be fully differential. They have a single output, which is referenced to system ground (Figure 4.8).

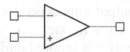

Figure 4.8
Single Ended Op Amp Schematic Symbol

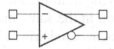

Figure 4.9
Fully Differential Op Amp Schematic Symbol

The op amp also has two power supply inputs, which are connected to bipolar power supplies (equal and opposite positive and negative potentials), or a single potential, with a positive supply and a ground connected to the power supply pins. These power supply pins are often omitted from the schematic symbol, when power supply connections are implied elsewhere on the schematic. Fully differential op amps add a second output (Figure 4.9).

The two outputs are fully differential; thus, the two outputs are called "positive output" and "negative output", using similar terminology to the two inputs. Like the inputs, they are differential. The output voltages will be equal, but opposite in polarity and referenced to the common mode operating point of the circuit.

4.4.2 How is the Second Output Used?

An op amp is used as a closed-loop device. From Chapter 2, the loops of simple gain stages are closed with R_f as shown in Figure 4.10.

Whether the single ended op amp is used in an inverting or a non-inverting mode, the loop is closed from the output to the inverting input.

4.4.3 Differential Gain Stages

You might be asking at this point: how is the loop closed on a fully differential op amp? It stands to reason that if there are two outputs, both of them have to be operated closed loop. Therefore, the equivalent way of closing the loop on a fully differential op amp is shown in Figure 4.11.

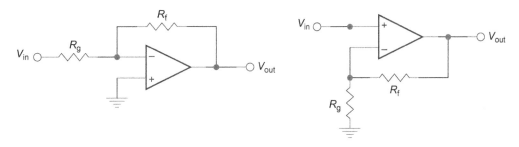

Figure 4.10
Closing the Loop on a Single-Ended Op Amp

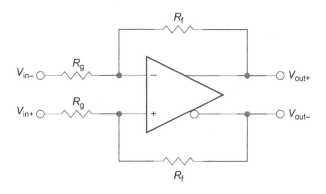

Figure 4.11
Closing the Loop on a Fully Differential Op Amp

Two identical feedback loops are required to close the loops for a fully differential op amp. If the loops are not matched, there can be significant second order harmonic distortion.

Note that for a fully differential op amp, each feedback loop is an inverting feedback loop. Both polarities of output are available, so terms like "inverting" and "non-inverting" are meaningless. Consider the inverting single-ended schematic in Figure 4.10. In this case, the loop goes from the (non-inverting) output to the inverting input, introducing a 180° phase shift. For the fully differential op amp, the top feedback loop has a 180° phase shift from the non-inverting output to the inverting input, and the bottom feedback loop has a 180° phase shift from the inverting output to the non-inverting input. Both feedback paths are therefore inverting. There is no "non-inverting" fully differential op amp gain circuit.

The gain of the differential stage is:

$$\frac{V_O}{V_I} = \frac{R_f}{R_g} \tag{4.6}$$

4.4.4 Single-Ended to Differential Conversion

The schematic shown in Figure 4.11 is a fully differential gain circuit. Fully differential applications, however, are some what limited. Very often the fully differential op amp is used to convert a single-ended signal to a differential signal, perhaps to connect to the differential input of an analog-to-digital converter.

The two configurations shown in Figure 4.12 are equivalent. At first glance, they look identical, but they are not. The difference is that in the left configuration, the inverting input is used for signal and the non-inverting input for reference. In the right configuration, the non-inverting input is used for signal and the inverting input for reference. They are functionally equivalent; either one will work.

The gain of the single-ended to differential stage is:

$$\frac{V_O}{V_I} = \frac{R_f}{R_g} \tag{4.7}$$

This is the same as Equation 4.6. The only difference between the configuration creating Equation 4.7 and the configuration creating Equation 4.6 is that one side of the input voltage is referenced to ground.

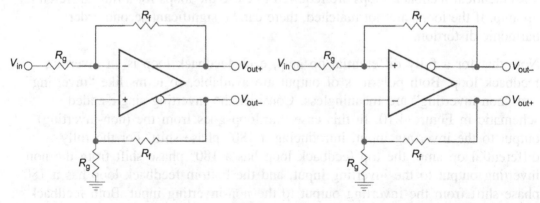

Figure 4.12
Single-Ended to Differential Conversion

The dynamics of the gain are sometimes best described pictorially. Figure 4.13 shows the relationship between V_{IN}, V_{OUT+}, and V_{OUT-} when $R_f = R_g$. What is going on here? Is the amplitude of the input, V_{IN}, twice that of the output? The gain is correct because the value of the differential gain $[(V_{OUT+}) - (V_{OUT-})]$ at any point in Figure 4.13 is equal to the amplitude of V_{IN}.

4.4.5 A New Function

Some fully differential op amps have an additional pin, V_{OCM}, which stands for "voltage output common mode (level)". The function of this pin can be either an input or an output, because its source is just a voltage divider off the power supply, but it is seldom used as an output. When it is used as an output, it will correspond to the common mode voltage about which the V_{OUT+} and V_{OUT-} outputs swing.

The most common use of the V_{OCM} pin is to set the output common mode level of the fully differential op amp. This is a very useful function, because it can be used to match the common mode point of a data converter to which the fully differential amplifier is connected. High-precision/high-speed data converters often employ differential inputs, and provide a reference output.

The schematic of Figure 4.14 is simplified, and does not show compensation, termination, or decoupling components for clarity. Nevertheless, it shows the basic concept. This is an important type of interface, and will be discussed further in Chapter 5.

4.5 Instrumentation Amplifier

An instrumentation amplifier is used to amplify a differential signal when both inputs need to be high impedance, usually because the source is high impedance. Figure 4.15 shows a common application, that of a strain gauge. A strain gauge consists of four resistive elements, one or more of which varies with applied mechanical stress. If the traditional differential amplifier of Figure 2.9 was used, the input impedances of the stage would load the strain gauge down, invalidating the measurement. The only way to combat this loading effect is with the three-amplifier implementation of Figure 4.15.

In this circuit configuration, both sources connect to the non-inverting input of two op amps. This impedance is very high, and if the op amps are identical, both impedances are very nearly equal.

When $R_7 = R_6$, $R_5 = R_2$, $R_1 = R_4$,

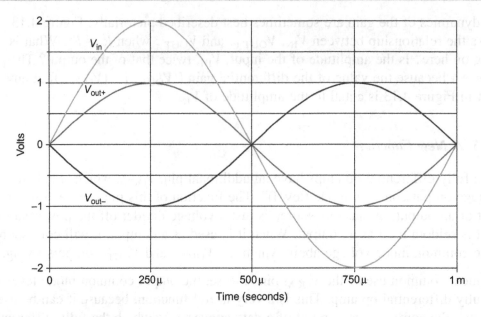

Figure 4.13

Relationship between V_{IN}, V_{OUT+}, and V_{OUT-}

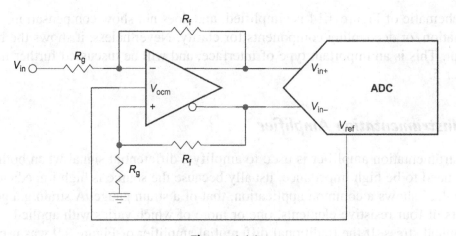

Figure 4.14

Using a Fully Differential Op Amp to Drive an Analog-to-Digital Converter

$$V_{OUT} = (V_{IN2} - V_{IN1})\left(\frac{2R_1}{R_3} + 1\right)\left(\frac{R_6}{R_7}\right) + V_{REF} \qquad (4.8)$$

This differential amplifier has the unique feature that the gain can be changed with only one resistor, R_3. Implementing instrumentation amplifiers, however, can

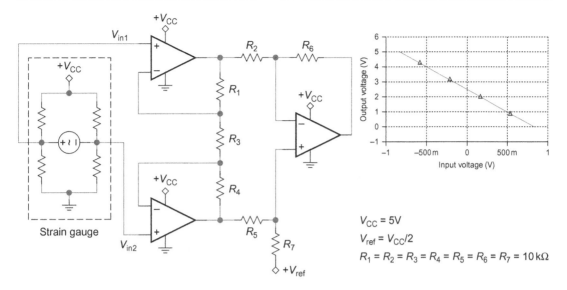

Figure 4.15
Instrumentation Amplifier

become troublesome, especially if board space and power are at a premium. Resistors should be matched with more precision than is expected from the circuit. Resistor mismatching increases distortion due to unequal gains, and it increases the common-mode voltage feedthrough. Resistors are hard to match, and matched resistor sets/arrays are expensive with long lead times. Many semiconductor manufacturers, therefore, have implemented the three op amp topology directly in silicon, freeing the user from the need to do so. Figure 4.16 shows a typical offering from a semiconductor company.

Resistors integrated on silicon can be matched to a high degree of precision, and board space is recovered because many of these ICs are implemented in 8-pin packages. Topologies inside the IC may differ from that shown above, so you are cautioned to read the data sheet carefully in order to design a circuit with the correct input impedance and gain characteristics. The output amplifier, for example, may not be a unity gain stage, and the ratio of R_f to R_g on the output stage may be 10 or even 100, making extremely high gains possible for very low-input signal levels. Also note that the topology above will still work even if R_g is omitted. Stage gain for Figure 4.16 would revert to 1.

4.6 Difference Amplifier

A variation on the instrumentation amplifier is the difference amplifier. These amps are used when the input voltage is larger than the supply voltage of the chip, and

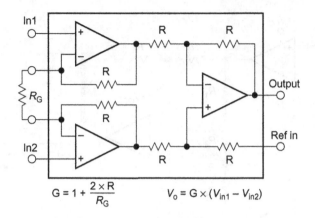

$$G = 1 + \frac{2 \times R}{R_G}$$

$$V_o = G \times (V_{in1} - V_{in2})$$

Figure 4.16
High-Precision Differential Amplifier

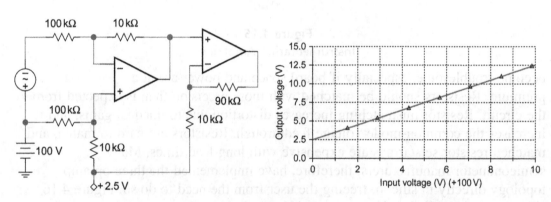

Figure 4.17
Difference Amplifier

are, by design, stable attenuators. They can be followed by gain stages so the net effect is to translate voltage levels that are difficult to deal with into a manageable voltage range. These most commonly involve voltage levels with high DC offsets. Figure 4.17 shows the use of a difference amplifier.

The difference amplifier of Figure 4.17 has an input of 0–10 V, with a DC offset of 100 V (making the actual input 100–110 V. This voltage is to be monitored at the output of the second op amp. The circuit operates off a power supply of 0–15 V.

The first stage performs the operation of removing the common mode 100 V DC offset. Because the input resistors are 100k and the gain resistors 10k, they effectively operate as voltage dividers, taking the input voltage down into the range where the op amp can handle the voltage. The input stage is a differential attenuator stage, which divides the input voltage by a factor of 10. In addition, a 2.5 V offset lifts the output off ground so the op amp V_{OL} does not clip the response. Therefore, the output of the first op amp is 2.5−3.5 V. The good news is that the 100 V DC offset has been eliminated. The bad news is that the voltage swing on top of the 100 V offset has been attenuated by a factor of 10:1. This does not take good advantage of the available voltage range of 0−15 V.

The second stage corrects this by amplifying the voltage by a factor of 10, while preserving the 2.5 V offset. This produces the output characteristic shown in Figure 4.17 where a swing of 100 V to 110 V on the input produces an output swing of 2.5 V to 12.5 V, easily within the V_{OH}/V_{OL} range of the output op amp, and ready to be interpreted by a data converter or meter calibrated to eliminate the 2.5 V offset.

Difference amplifiers are most commonly used as "high side current monitors" for power supplies. Figure 4.18 illustrates this application. This figure is actually a minor variation of Figure 4.17. The schematic has been rearranged slightly to show the components commonly integrated into an IC (the dashed line). The 100 V DC offset is the power supply to be monitored, and the signal source that was above it has been replaced with a current sense resistor, R_S. The output of the sense resistor is connected to a load R_L, which will be assumed to be 99.9Ω.

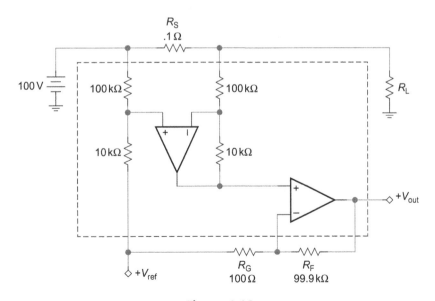

Figure 4.18
High Side Current Monitor

You might recognize at this point that R_S and R_L form a voltage divider. Because R_S is such a small value, it does not contribute much voltage drop compared to R_L. You should also exercise caution: R_S will probably need to be a larger wattage resistor if R_L is drawing appreciable current.

In this case the total load on the 100 V power supply is 100Ω, and therefore the current through R_S and R_L is 1 A. The wattage of R_S is therefore 0.1 W. The wattage in R_L is your responsibility, and is assumed to be distributed among many active components in an application circuit. The voltage drop across R_S is limited to 0.1 V, which leaves 99.9 V available to the load, a drop of only 0.1%. $+V_{REF}$ is still $+2.5$ V. The second amplifier is now operated at a gain of 1000, which will give an output at V_{OUT} of 12.5 V for the fully loaded power supply operating at 1 A, and V_{OUT} will be 2.5 V for an open circuit. R_f could be made a standard value of 100 kΩ with only a 0.1% measurement error.

An example difference amplifier is shown in Figure 4.19. This particular model is the AD628 or INA146; these are functional but not pin-equivalent devices, and happen to be applicable to the design examples above. These devices have made provision for an RC low-pass filter to reduce noise, which should not be confused with a compensation network. Other models in the product lines from semiconductor manufacturers are designed for different levels of attenuation

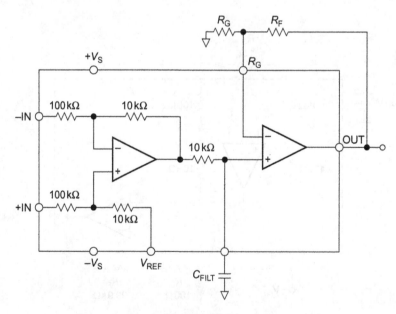

Figure 4.19
Commercial Difference Amplifier

(allowing more or less common mode voltage). Some may omit the output amplifier. You should read the data sheet carefully to determine what device is best for your application.

So, why buy a special difference amplifier IC instead of just implementing with ordinary op amps? Referring back to the discussion in Section 4.2 for undercompensated op amps, op amp attenuators are inherently unstable; this applies to differential stages as well as simple inverting stages. Think of the differential stage in Figure 4.17 as a simple inverting attenuator with an offset applied to the non-inverting input: it is still an inverting attenuator and therefore unstable. The input amplifier in IC difference amplifiers has been designed to be gain of 0.1 stable, while the output amplifier is designed for high gains with low offsets, which are two very different stability and performance criteria. You can always keep this in your arsenal of design tricks if you really want to design an inverting op amp attenuator.

4.7 Buffer Amplifiers

The op amps discussed until this point have had something in common. They are somewhat limited in the amount of power that they can drive with their output. In general, VFAs can drive a 600Ω load fairly well, but are not designed for lower impedances. CFAs, on the other hand, are often designed with very robust output stages. In fact, a whole class of line driver CFAs was designed for digital subscriber line (DSL) applications. Unfortunately, DSL is losing traction in favor of cable and fiber to home networking. The high-output power op amps remain for now, but may become obsolete as time goes by.

Fortunately, another class of amplifiers is available: buffer amplifiers. Buffer amplifiers can be thought of as integrated unity gain buffers: hook up a power supply, bypass it properly, apply an input signal, and connect the output to the load. No resistors required! This is very slick way to drive heavy loads such as long cables and audio loads.

Buffer amplifiers, being active devices, will add their own characteristics to the signal. Fortunately, there is a way to partially cancel undesirable buffer amplifier effects: put them inside the loop!

The circuit of Figure 4.20 assumes a unity gain, non-inverting buffer amplifier. The performance of this "hybrid amplifier" circuit can actually be better than just using a single amplifier, because the precision input amplifier is free from supplying load current to V_{OUT}, and therefore will not heat up as much and drift. Figure 4.20 shows split-supply operation. Single-supply operation is also possible as long as you keep

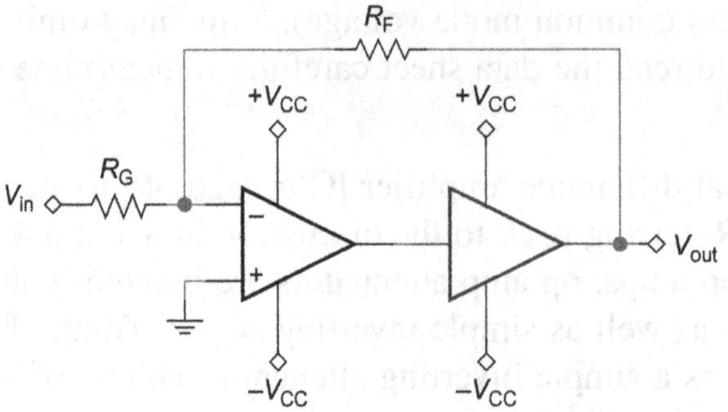

Figure 4.20
A Better Way to Use a Buffer Amplifier

the V_{OH} and V_{OL} specifications of the buffer in mind; they may also vary with the load. There is no reason why the buffer amplifier cannot be operated off a different (higher) set of rails, but you must be careful that the input amplifier never sees a voltage greater than its normal operating input range. This means there must be an appreciable gain in the circuit so that R_f and R_g form a voltage divider on the output voltage keeping the inverting input of the hybrid stage in the normal operating region.

Power stages have sources of instability that have not been discussed here, associated with their heavy loads. You should carefully follow data sheet instructions for bypassing, maximum inductance and capacitance on inputs and outputs. There may also be recommendations for snubber networks to suppress unwanted high-frequency oscillation. Suffice it to say that dealing with heavy loads is a complex task and should be taken seriously.

The buffer amplifiers can be replaced with non-inverting op amp circuits. CFAs often have an output stage more robust than that for VFAs, and as long as you are careful to observe stability recommendations for CFAs, they will function well as buffer amplifiers. Power VFAs are also available, some of which can drive several amps of load current at high voltage levels. You must pay special attention to the data sheet characteristics and recommendations for these devices, and they often require a heatsink to operate at their recommended loads.

If even more output power is required, it is possible to parallel buffer amplifiers (Figure 4.21). In order to insure proper current sharing, small values of output resistor (R_{O1} and R_{O2}), usually between 1 and 5 Ω, are placed at each amplifier output. This causes a decrease in output voltage swing, because the series resistors act as a voltage divider with the load. But without the series resistors, the buffers

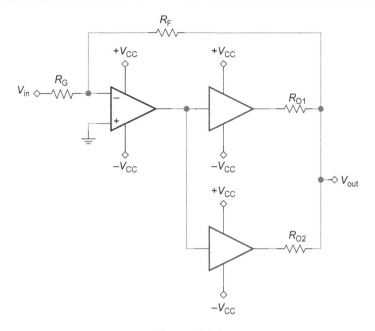

Figure 4.21
Paralleling Buffer Amplifiers

would tend to drive each other into oscillation. Usually a bit of experimentation is required to find the correct value. You should also make sure the resistors are of the correct wattage.

There is no theoretical limit on how many buffers can be placed in parallel; however, printed circuit board parasitics and parasitics associated with the amplifiers themselves usually limit the number to a handful of devices.

The hybrid amplifier will be discussed in more detail in Chapter 11.

4.8 Other Types of Op Amps

The list of op amp varieties could go on and on: chopper amplifier, Norton amplifiers, variable gain amplifiers, to name a few. Most of these types of amplifiers are specialized enough, and require detailed explanations beyond the scope of this volume. If you need one of them, chances are that you are already sufficiently well versed in op amp design that you will understand the reasons why you need it. There are many good reference works on each of the types I have mentioned, as well as the dozens of other types.

CHAPTER 5

Interfacing a Transducer to an Analog-to-Digital Converter

5.1 Introduction

One of the most common questions asked by customers is: "Which op amp should I use with a [substitute a part number] ADC?"

The question of which data converter to use with any given data converter is one that makes marketing people cringe, because the answer touches on different technical areas that they are probably not conversant in. It is also an uncomfortable question for applications support engineers, because the customer may not have a clear idea of all the issues and trade-offs involved.

The process of selecting an op amp to drive a data converter is an exercise in weeding out those op amps that will clearly not work. The subset of op amps left at the end of an elimination process is much more manageable.

Op amp manufacturers have simultaneously made things easier, and harder, for you. Some manufacturer websites have sections targeted to major product groups, and op amps appropriate for that product group will pop to the top of the list. If that is not the case, the number of op amp listings is sometimes vast, and it may not be evident which op amp has been optimized for a product group. But each new op amp manufactured is the result of a product group meeting, where a group of integrated circuit (IC) designers, managers, and marketing people have decided that optimizing four or five parameters will allow them to sell an op amp to manufacturers of a particular type of product. Your job is to discover that op amp, and this procedure will lead you in that direction.

Make no mistake, the list of questions below is daunting. However, these are important questions that must be answered, for the most part, before doing the design. You have probably already answered most of them without realizing it; or the answer is so self-evident that the question was not asked. The answer to one question can automatically eliminate many other questions.

Op Amps for Everyone.
DOI: http://dx.doi.org/10.1016/B978-0-12-391495-8.00005-2

I have divided the questions into broad categories:

- The system as a whole: understanding what the product is supposed to do can lead you in useful directions.
- The power supply: power supply voltages in electronics, especially portable devices, have been trending downwards. Customers want lower and lower battery voltage, but with no compromise in performance. The first Regency transistor radio, for example, used a 22.5 V battery. My childhood transistor radio used a 9 V battery. My current portable radio runs off a single 1.5 V battery, yet offers much better performance than the older radios.
- Input signal characteristics: the type of signal being input to the stage can greatly influence the choice of op amp. Signals that are audio or radiofrequency (RF) can be AC coupled, and DC specifications of the op amp are not important. Some sensors, on the other hand, are almost entirely DC in nature, so AC performance of the op amp is not important.
- Analog-to-digital converter (ADC) characteristics: a single-ended data converter will use single-ended op amps, while a fully differential data converter will require fully differential drive.
- Op amp characteristics: sometimes packaging, temperature range, or other considerations influence which op amp is used.

And finally, I will offer some hints about how to properly drive a fully differential ADC.

Since there are so many questions, I suggest that they be discussed in a "kick-off" meeting environment, where all designers associated with the system can freely input ideas. None of the questions is intended to be a "sticking point" or closed door, merely a suggested consideration that might affect system design. I know how meetings go, how one point can grow into a half-hour debate, but that is not the intention. Each of these questions should be simple, the answer self-evident when asked of the right person, so system-level discussions should go very quickly.

5.2 System Information

The overall characteristics of the system often yield valuable information. A clear understanding of the product and its function is imperative to design success, and can start the process of weeding out unsuitable parts.

- Exactly what is the end equipment and its application? Different systems have different requirements.
  Examples:
  - A transducer interface design requires DC accuracy and leads to DC accurate op amps.

- Wireless communication systems will require high-speed op amps with good RF specifications.
- In general terms, what is the function of this signal-acquisition chain in the system? Where does the input signal come from and what happens to it once it is digitized?

  Examples:
  - A slowly changing DC signal can utilize slow op amps optimized for DC accuracy.
  - An RF system will be AC coupled and must run at least as quickly as the Nyquist frequency.
  - Audio systems require op amps with low levels of total harmonic distortion.
- How many signal-acquisition chains are used in the product? Channel density can influence system design in numerous ways, including space constraints, thermal requirements, and amplifier channel density per package.

  Example:
  - A medical ultrasound device can have 100 or more channels, leading to challenges of component count, board size, power consumption, and heating; so you will want quad op amps, low in power consumption, in small packages, probably operating off a low voltage.
- Where will the system be used? What temperature conditions will the system operate in?

  Examples:
  - Military, space, downhole, and geothermal are all applications where the signal chain will be subjected to extremes of temperature. They will require high-reliability components that are probably already on an approved list.
  - Consumer electronics will probably be subjected to nothing more severe than a hot car dashboard or an overnight freeze; therefore the op amps will not have to be nearly as temperature tolerant as the categories listed above.

5.3 Power Supply Information

Power supply rails can quickly rule out op amps. This is similar to clothes shopping: the style may be desirable, but if the size does not fit, the style is useless. So a wise shopper finds the options in the size first, before becoming attached to a style. Similarly, power supply information is collected first (Figure 5.1), because it will narrow op amp choices. Which power supply voltages are present in the system? Are ± 15 V supplies available? ± 5 V? No negative supplies at all? Only low voltage, $+3$ V from batteries or even $+1.5$ V? An op amp with fantastic specifications at ± 15 V may not operate at all from $+3$ V; remember the V_{OL} and V_{OH} specifications of the op amp. Therefore, if $+3$ V is all that is available in the

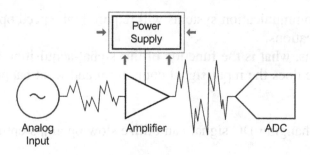

Figure 5.1
Focusing on the Power Supply Characteristics

system, you can rule out all but single-supply, rail-to-rail devices, and must design within their limitations.

Examples:
- Legacy analog systems almost always have ± 15 V rails. Upgrading them with newer generation op amps requires that the new op amps can operate at these voltages.
- Many high-end data acquisition systems have standardized on ± 5 V rails. Many devices are offered, perhaps the majority of op amps in this voltage rating. They may not be able to operate off ± 15 V rails.
- Portable equipment (battery operated) tends to operate off batteries that will provide multiples of 1.5 V. Even systems that use button cells have multiples of 3 V. Lithium ion battery systems will have multiples of 4 V. Of course, the ultimate in lightweight, small, portable electronics will operate off a single 1.5 V or 3 V battery. But because the cell voltage drops as the battery is depleted, these systems must operate off a voltage lower than 1.5 V, sometimes as little as 0.8 V, creating an extreme limitation on the number of suitable op amps.
- Is a precision reference available in the system? In single-supply systems, it is important to supply a virtual ground to the op amp circuitry.

Example:
- Higher end data acquisition systems tend to use ADCs with built-in references. You should use this reference if at all possible.

5.4 Input Signal Characteristics

Understanding the input source (Figure 5.2) is key to properly designing the interface circuitry between the source and the ADC.

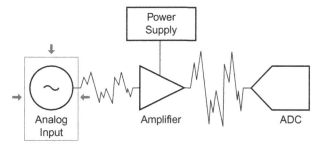

Figure 5.2
Focusing on the Input Signal

- What is the output amplitude range of the source? This information is used in conjunction with the cases in Chapter 3. The source determines $V_{in\ minimum}$ and $V_{in\ maximum}$.
- Does the source produce a current output? This requires a different topology.
  Example:
  - Some temperature sensors.
- Is the signal source output single-ended or differential? A differential input signal may lend itself to an instrumentation amplifier rather than a single-ended op amp.
  Examples:
  - 600Ω balanced audio
  - pressure transducers (strain gauges).
- What is the output impedance of the signal source? Very high impedance sources require even higher impedance amplifiers. This will definitely dictate a non-inverting op amp topology, but even that may not suffice. It may require extremely high impedance junction field effect transistor (J-FET) op amps or instrumentation amplifiers.
  Examples:
  - photomultiplier tubes
  - pressure transducers (strain gauges).

5.5 Analog-to-Digital Converter Characteristics

Now that the power supply and input signal have been defined, it is time to focus on the device that the op amp will drive: the ADC (Figure 5.3).

- What is the full-scale input range of the data converter? The ADC input low and high voltages, along with values from the input signal section above, determine the "case" of Chapter 3.

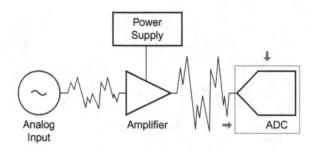

Figure 5.3
Focusing on the Analog-to-Digital Converter

- Will the data converter be used with single-ended or differential inputs? Typically, most high-performance data converters have differential inputs and require their use for optimum performance. You will, however, have to convert a single-ended signal to a fully differential signal to obtain maximum performance from the ADC.
- What is the desired resolution and effective number of bits? A 14-bit converter will not effectively yield 14 bits. The true resolution will probably be closer to 12 or 13. If 14-bit performance is really desired, ask whether a 16-bit converter can be substituted. Often there is a family of similar data converters, and a higher resolution converter may be pin-for-pin compatible.
- What is the desired sampling rate? Often, people assume that a data converter is going to be used at its maximum sampling rate, but sometimes this comes at the cost of accuracy. For example, an 80 mega-sample per second (MSps) converter might be given a sampling frequency of 60 MSps to achieve greater accuracy.
- Are there any compensation requirements for the input of the data converter? Normally, a small RC filter is required at the input of the data converter to compensate for its capacitive input. These components are specified in the converter data sheet, and should be included as part of the interface. Otherwise, the op amp interface circuit may exhibit instability.

5.6 Interface Characteristics

By now you have narrowed down the potential choices for an op amp by its supply voltages. You know what "case" your interface requires, which will give you a rough idea of the schematic of the interface. You also know whether you need a single-ended to fully differential conversion stage. But there are some other pieces of information you need in order to flesh out the complete input signal to ADC input interface (Figure 5.4).

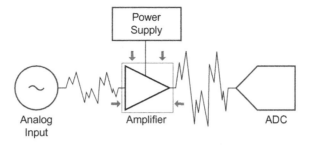

Figure 5.4
Focusing on the Operational Amplifiers

- Is the signal DC accurate, high speed, audio, or RF? You can consult Appendix A for a reference on which op amp parameters are important for different types of applications.
  Examples:
  - Pressure and temperature transducer circuits are almost exclusively DC accurate. You need an op amp with good DC specifications.
  - High-speed systems may go hand in hand with RF, or just require fast op amps because of filter constraints. Consult Section 6.4 (Chapter 6) for high-speed filter considerations; you may be surprised just how fast an op amp needs to be to operate in an active filter circuit.
  - Audio requires op amps with low noise in the audio bandwidth. Some low-noise, high-speed op amps may have bad noise specifications at low frequency.
  - RF applications require a completely different set of specifications. Consult Chapter 7 and Appendix A.
- Do you need to filter the signal?
  Examples:
  - Reduce high-frequency noise (low-pass filter).
  - Reduce low-frequency noise (high-pass filter).
  - Detect a single frequency (bandpass filter).
  - Reject an interfering frequency (notch filter).
- These filters will be discussed in the next chapter.
- Are there specific requirements for the package of the amplifiers?
  Examples:
  - Small, surface mount? Or does package size not matter?
  - Does it have to be a high-reliability ceramic package?

Now, you have a really good idea of which op amps you cannot use, and hopefully a small list of the ones you can. You should take that small list and evaluate them based on their suitability. Perhaps make use of one of the free simulation programs listed in Chapter 12, or even order them and try them in your prototype circuit. Do not be afraid to experiment. Options are sometimes good, so if half a dozen parts meet your requirements, be glad because you can put them as alternatives on your bill of materials!

5.7 Architectural Decisions

Even with your op amp selection(s) made, the job is still not quite done.

- As I mentioned above, you may need a compensation network. This network will also form a low-pass filter, but fortunately the low-pass characteristic is above the operating frequency of the ADC.
- I would put gain stages first, then filter stages. In the case of a lot of gain, I would use two or more gain stages.
- Of course, if you have a fully differential ADC and a single-ended signal source, you need to do a conversion. I will spend the rest of this section on this interface.

Figure 5.5 shows a typical single-ended to fully differential interface circuit. The input signal is referenced to ground, while the common mode operating point of the op amp interface is set by the data converter. The op amp interface can be run off a single supply. DC blocking capacitors C_1 and C_2 prevent the common mode point from being affected by the input signal. Response down to DC, of course, has been sacrificed, but this may be acceptable in most applications.

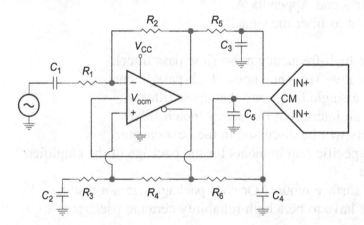

Figure 5.5
Single-Ended to Fully Differential AC Coupled Interface

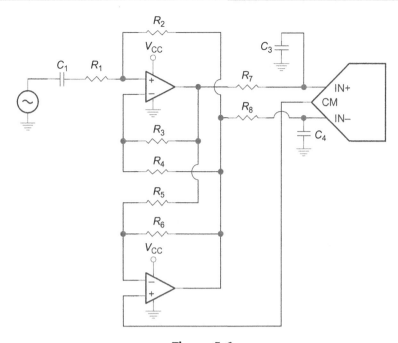

Figure 5.6
Single-Ended to Fully Differential AC Coupled Interface

R_5, C_3, R_6, and C_4 form the compensation networks defined by the ADC data sheet. If this circuit requires only gain and no filtering, it may be possible to use the fully differential op amp as the entire interface. If filtering is required, it can be done single-ended and input to C_1.

It is not absolutely necessary to use a fully differential amplifier to drive a differential ADC. Figure 5.6 shows the preferred way to convert single-ended signals to differential without a transformer. While this circuit looks a bit unusual, the strategy is to equalize the delay for IN+ and IN− by forcing each phase of the signal to go through two op amps before being applied to the inputs. This may not be intuitive at first glance. Each amplifier, though, is in the feedback loop for the other. Think of this as an inverting op amp circuit: gain is adjusted by changing R_1 (corresponding to R_g), and R_2–R_6 are equal values (corresponding to R_f). This circuit is referenced to the ADC reference in the bottom op amp.

5.8 Conclusions

Selecting the right op amp(s) for an interface between a sensor and a data converter can be a daunting challenge. Selecting the op amps is more of an exercise in weeding out unsuitable devices, at least in the early stages. You can use a systematic approach to the problem, and tackle the issues and questions one section of circuitry at a time. You can narrow down the op amp choices by:

• analyzing the nature of the system
• knowing the power supply voltages in the system
• knowing the input signal characteristics
• knowing the ADC characteristics

and, finally, by:

• knowing something about the op amps themselves
• knowing how to interface with the ADC.

These questions may be formidable, but they are essential if you want to develop a working interface. Take the time to ask them and understand the implications of the answers.

CHAPTER 6

Active Filter Design Techniques

6.1 Introduction

What is a filter? According to Webster, it is "a device that passes electric signals at certain frequencies or frequency ranges while preventing the passage of others".

Filter circuits are used in a wide variety of applications. In the field of telecommunications, bandpass filters are used in the audio frequency range (20 Hz to 20 kHz) for modems and speech processing. High-frequency bandpass filters (several hundred megahertz) are used for channel selection in telephone central offices. Data acquisition systems usually require antialiasing low-pass filters as well as low-pass noise filters in their preceding signal conditioning stages. System power supplies often use band-rejection filters to suppress the 60 Hz line-frequency and high-frequency transients.

Entire books have been written on the subject of filter design. As a young engineer at the beginning of my career, I purchased several of them from a nearby university book store. I was completely dissatisfied with them from the start: they took a mathematical/theoretical approach, and therefore were cumbersome to use when actually designing a filter. What I needed was a fast, practical design approach. After many years in this profession, I wrote my own book, one I could finally be satisfied with. I have managed to come up with manageable approaches to different types of filters, and will present them here. I have selected topologies for each type of filter that will minimize the number of op amps used.

In over 30 years as an engineer, I have never encountered a set of design requirements that said to use more op amps than are necessary, consume as much power as you deem necessary, take up extra board space with more complex circuit topologies, and decrease reliability by using more parts. Therefore, this chapter will not present topologies you will see in other books such as biquad filters, for the reason that they use three op amps when a single op amp can do the same job. I will give a slight nod to the biquad filter for flexibility and the ability to output more than one response at a time. But with the possible exception of test equipment, this advantage comes at too high a cost.

Op Amps for Everyone.

DOI: http://dx.doi.org/10.1016/B978-0-12-391495-8.00006-4

I will first discuss different methods of filter design, and why I think they are difficult and cumbersome. After that, I will discuss my method. At the end of the chapter, as I did for gain and offset, I will give design aids and printed circuit board (PCB) layouts. Some semiconductor manufacturers have posted free filter design utilities that I will discuss in Chapter 12.

6.2 The Transfer Equation Method

The transfer equation method of designing active filters is a mathematical approach that has been championed in textbooks for decades. I certainly do not belittle it: if you are comfortable in the mathematical arena, these methods can produce a very satisfactory result. The transfer equation method is also the most general method: it will work for any filter topology and therefore is ideal for esoteric cases. This book will not use the transfer equation method for reasons I will state below.

I have condensed a single case of filter design using the transfer equation method from Thomas Kugelstadt's excellent write-up in *Op Amps for Everyone*, the ancestral work to this volume. It remains one of the most concise and easily understood treatises on the subject of transfer equation filter design ever written, and I highly recommend it for proponents of this filter design method. Without further introduction, here is an example of a transfer equation design of a two-pole, unity-gain Sallen−Key low-pass filter.

Equation 6.1 represents a general second order low-pass filter. The transfer function of a single stage is:

$$A_i(S) = \frac{A_0}{(1 + a_i S + b_i S^2)} \tag{6.1}$$

The unity-gain Sallen−Key topology in Figure 6.1 is usually applied in filter designs with high gain accuracy, unity gain, and low Q ($Q < 3$).

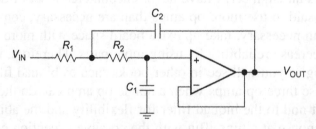

Figure 6.1
Unity-Gain Sallen−Key Low-Pass Filter

The transfer function of the circuit in Figure 6.1 is:

$$A(S) = \frac{1}{1 + \omega_c C_1 (R_1 + R_2)S + \omega_c^2 R_1 R_2 C_1 C_2 S^2} \tag{6.2}$$

The coefficient comparison between this transfer function and Equation 6.2 yields:

$$\begin{aligned} A_0 &= 1 \\ a_1 &= \omega_c C_1 (R_1 + R_2) \\ b_1 &= \omega_c^2 R_1 R_2 C_1 C_2 \end{aligned} \tag{6.3}$$

Given C_1 and C_2, the resistor values for R_1 and R_2 are calculated through:

$$R_{1,2} = \frac{a_1 C_2 \mp \sqrt{a_1^2 C_2^2 - 4 b_1 C_1 C_2}}{4 \pi f_c C_1 C_2} \tag{6.4}$$

In order to obtain real values under the square root, C_2 must satisfy the following condition:

$$C_2 \geq C_1 \frac{4 b_1}{a_1^2} \tag{6.5}$$

Example: Second Order Unity-Gain Tschebyscheff Low-Pass Filter

The task is to design a second order unity-gain Tschebyscheff low-pass filter with a corner frequency of $F_C = 3$ kHz and a 3 dB passband ripple.

The coefficients a_1 and b_1 are obtained from Table 6.1:

$$\begin{aligned} a_1 &= 1.0650 \\ b_1 &= 1.9305 \end{aligned} \tag{6.6}$$

Specifying C_1 as 22 nF yields a C_2 of:

$$C_2 \geq C_1 \frac{4 b_1}{a_1^2} = 22 \cdot 10^{-9} \text{nF} \cdot \frac{4 \cdot 1.9305}{1.065^2} \cong 150 \text{ nF} \tag{6.7}$$

Table 6.1: Second Order Filter Coefficients

Second Order	Bessel	Butterworth	3 dB Tschebyscheff
a_1	1.3617	1.4142	1.065
b_1	0.618	1	1.9305
Q	0.58	0.71	1.3
R_4/R_3	0.268	0.568	0.234

Inserting a_1 and b_1 into the resistor equation for $R_{1,2}$ results in:

$$R_1 = \frac{1.065 \cdot 150 \cdot 10^{-9} - \sqrt{\left(1.065 \cdot 150 \cdot 10^{-9}\right)^2 - 4 \cdot 1.9305 \cdot 22 \cdot 10^{-9} \cdot 150 \cdot 10^{-9}}}{4\pi \cdot 3 \cdot 10^3 \cdot 22 \cdot 10^{-9} \cdot 150 \cdot 10^{-9}} = 1.26\,k\Omega$$

(6.8)

and

$$R_2 = \frac{1.065 \cdot 150 \cdot 10^{-9} - \sqrt{\left(1.065 \cdot 150 \cdot 10^{-9}\right)^2 - 4 \cdot 1.9305 \cdot 22 \cdot 10^{-9} \cdot 150 \cdot 10^{-9}}}{4\pi \cdot 3 \cdot 10^3 \cdot 22 \cdot 10^{-9} \cdot 150 \cdot 10^{-9}} = 1.30\,k\Omega$$

(6.9)

with the final circuit shown in Figure 6.2.

So, why not just proceed to use this method for filter design?

- The transfer equation method works well for simple filters such as the one above. As filters become more complex, with more poles, and different topologies for example, the math quickly becomes unwieldy. Since filter designers are not necessarily mathematicians, forcing them to derive transfer equations for each circuit is not an efficient means of design. Transfer equation derivation for filter design easily fills an entire book, and is beyond the scope of this volume.
- If you have the transfer equation of a given filter topology, you can use Mathcad or Matlab to generate a filter design solution. Transfer equations exist for common filter topologies, and this may be preferable to you. However, transfer equations are not always available for more exotic filter topologies or multiple pole filters.
- The transfer equation method does not result in real-world component values. There are too many textbooks on the subject that give precise values of

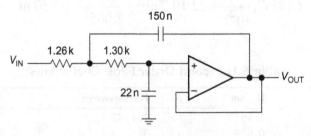

Figure 6.2
Second Order Unity-Gain Tschebyscheff Low-Pass with 3 dB Ripple

capacitors and resistors, and then leave you with no guidance about how to pick real values, in particular how to decide the scale of the components.
- The example above relied on a look-up table for filter coefficients. If you are going to use a table anyway, why not use a table for everything? There are textbooks that give filter designs in terms of R and C values on a graph of frequency and Q, and these yield fairly good designs quickly.

6.3 Fast, Practical Filter Design

The previous section introduced a rigorous theoretical approach to filter design. Many designers prefer this method because it gives the most flexibility in filter design. Years of customer support, however, have revealed that the vast majority of designers want a simple filter with the minimum of effort and development time. This section presents a pragmatic, simple design methodology that will allow you to implement all but the most complex filters rapidly, and have a reasonable expectation that they will be producible.

Some compromises have been made in order to keep this quick filter design process simple:

- A single op amp topology has been chosen for each filter type. I have picked the simplest (least number of op amps) approach. I do not cover multiple op amp topology such as biquad.
- Filter circuit responses are all Butterworth. I do not cover Tschebyscheff or Bessel responses.
- Filter gains are unity (gain of 1) except for band pass.

If you cannot work within those limitations, consult another source. I have found that these limitations are not extreme: 90% of filter design applications can be realized with these circuits. When combined with the gain/offset techniques of a previous chapter, these filter circuits provide a reasonable signal chain solution for most applications.

To design a filter, some things must be known in advance:

- the frequencies that need to be passed, and those that need to be rejected
- a transition frequency, the point at which the filter starts to work; or a center frequency around which the filter is symmetrical
- an initial capacitor value: pick one somewhere from 100 pF for high frequencies to 0.1 μF for low frequencies; if the resulting resistor values are too large or too small, pick another capacitor value.

6.3.1 Picking the Response

For the beginner, the filter responses will be presented pictorially. The area shaded in blue represents the frequencies that will be passed, and the area in white the frequencies that will be rejected. Do not be concerned with the exact frequency yet; that will be taken care of in the following sections. Look at the responses in Figures 6.3–6.7, and pick one where the desired frequencies are in the shaded area and the rejected frequencies in the unshaded area.

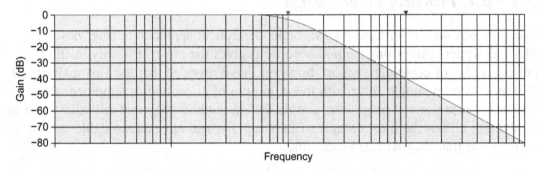

Figure 6.3
Low-Pass Response — Go to Section 6.3.2

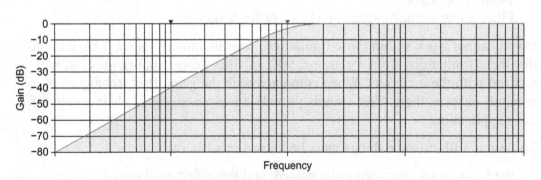

Figure 6.4
High-Pass Response — Go to Section 6.3.3

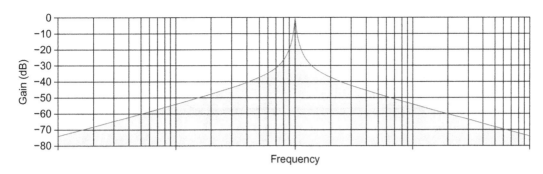

Figure 6.5
Narrow (Single-Frequency) Band Pass — Go to Section 6.3.4

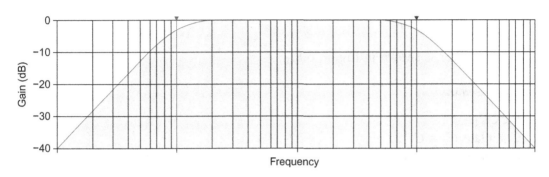

Figure 6.6
Wide Band Pass — Go to Section 6.3.6

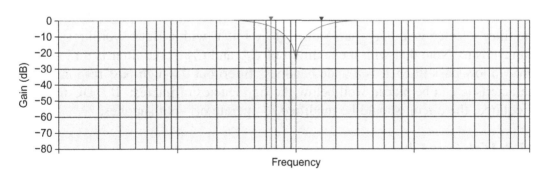

Figure 6.7
Single-Frequency Notch Filter

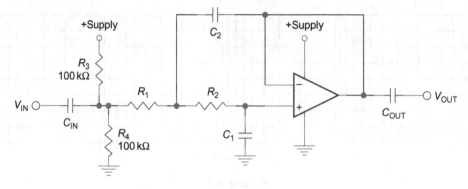

Figure 6.8
Low-Pass Filter

6.3.2 Low-Pass Filter

A low-pass filter is shown in Figure 6.8.

Design Procedure

- Pick C_1.
- Calculate $C_2 = C_1 * 2$.
- Calculate R_1 and R_2: $\dfrac{1}{2\sqrt{2}*\pi*C_1*\text{Frequency}}$.
- Calculate $C_{\text{IN}} = C_{\text{OUT}} = 100$ to 1000 times C_1 (not critical).
- DONE!

Digging Deeper

The filter selected is a unity-gain Sallen–Key Filter, with a Butterworth response characteristic. Note that with the addition of C_{IN} and C_{OUT}, the filter is no longer purely a low-pass filter. It is a wide bandpass filter, but the high-pass response characteristic can be placed well below the frequencies of interest. If a DC response is required, the circuit should be modified to operate off split supplies.

6.3.3 High-Pass Filter

A high-pass filter is shown in Figure 6.9.

Design Procedure

- Pick $C_1 = C_2$.
- Calculate R_1: $\dfrac{1}{\sqrt{2}*\pi*C_1*\text{Frequency}}$.

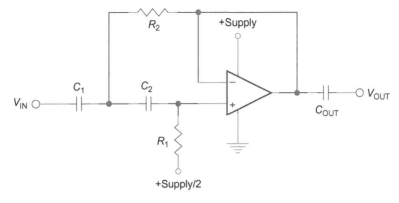

Figure 6.9
High-Pass Filter

- Calculate R_2: $\dfrac{1}{2\sqrt{2}*\pi*C_1*\text{Frequency}}$.
- Calculate $C_{\text{OUT}} = 100$ to 1000 times C_1 (not critical).
- DONE!

Digging Deeper

The filter selected is a unity-gain Sallen–Key filter, with a Butterworth response characteristic. Just as was the case with the low-pass filter, there is no such thing as an active high-pass filter, but for a different reason. The gain/bandwidth product of the op amp used will ultimately produce a low-pass response characteristic, making this a wide bandpass filter. It is your responsibility to choose an op amp with a frequency limit well above the bandwidth of interest.

6.3.4 Narrow (Single-Frequency) Bandpass Filter

A narrow bandpass filter is shown in Figure 6.10.

Design Procedure

- Pick $C_1 = C_2$.
- Calculate $R_1 = R_4$: $\dfrac{1}{2*\pi*C_1*\text{Frequency}}$.
- Calculate $R_3 = 19*R_1$.
- Calculate $R_2 = \dfrac{R_1}{19}$.
- Calculate $C_{\text{IN}} = C_{\text{OUT}} = 100$ to 1000 times C_1 (not critical).
- DONE!

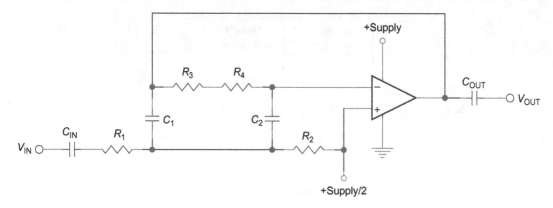

Figure 6.10
Narrow Bandpass Filter

Digging Deeper

The filter selected is a modified Deliyannis filter. A Deliyannis filter is a special case of the multiple-feedback (MFB) bandpass configuration, one that is very stable and relatively insensitive to component variation. The Q is set at 10, which also locks the gain at 10, as the two are related by the expression:

$$\frac{R_3 + R_4}{2 \cdot R_1} = Q = \text{Gain} \tag{6.10}$$

A higher Q was not selected, because the op amp gain bandwidth product can be easily reached, even with a gain of 20 dB. At least 40 dB of headroom should be allowed above the center frequency peak. The op amp slew rate should also be sufficient to allow the waveform at the center frequency to swing to the amplitude required.

6.3.5 Wide Bandpass Filter

A wide bandpass filter is shown in Figure 6.11.

Design Procedure

- Go to Section 6.3.3, and design a high-pass filter for the low end of the band.
- Go to Section 6.3.2, and design a low-pass filter for the high end of the band.
- Calculate $C_{IN} = C_{OUT} = 100$ to 1000 times C_1 in the low-pass filter section (not critical).
- DONE!

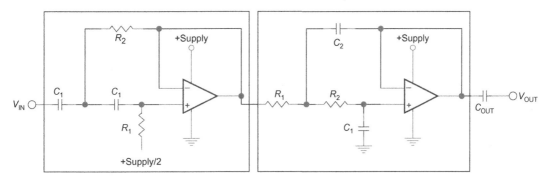

Figure 6.11
Wide Bandpass Filter

Digging Deeper

This is nothing more than cascaded Sallen–Key high-pass and low-pass filters. The high pass comes first, so noise from it will be low passed.

Digging Deeper: Narrow vs. Wide Bandpass Filter

At what point is it better to implement a bandpass filter as a narrow/single-frequency filter rather than as a wide band pass? At high Q values, the single-frequency band pass is clearly the better choice. However, as Q values decrease, the difference begins to blur. What can be a very sharp peak at resonance erodes to a single-pole roll-off on the low end, and single-pole roll-off on the high end. This results in a lot of unwanted energy in the stop bands.

For Q values of 0.1 (and below) and 0.2, the best implementation is high pass cascaded with low pass. The two implementations have almost an identical pass band response for a Q of 0.5. You are presented with a choice: use a bandpass filter (which can be implemented with a single op amp) to save money, or use a cascaded approach that has better rejection in the stop bands. As the Q becomes higher and higher, however, the responses of two separate stages begin to interact, destroying the amplitude of the signal. A good rule of thumb is that the start and ending frequencies of a wide bandpass filter should be at least a factor of five different.

6.3.6 Notch (Single-Frequency Rejection) Filter

Design Procedure

- Pick C_o.
- Calculate R_o: $\dfrac{1}{2*\pi*C_1*\text{Frequency}}$.

- Calculate $R_Q = 20 * R_o$.
- If you do not want to tune, replace R_{o_low} and R_{o_adj} with R_o. If tuning of the center frequency is desired, make a pot part of the value of R_o by going down one standard resistor value, and making sure the pot covers the range of center frequencies. If you do the job right, you can fine-tune the center frequency while preserving the depth of the notch. The nice thing about this topology is that you will get a very deep notch — somewhere!
- DONE!

Digging Deeper

This is the Fliege filter topology, set to a Q of 10. The Q can be adjusted independently from the center frequency by changing R_Q. Q is related to the center frequency set resistor by the following:

$$R_Q = 2 * Q * R_o$$

The Fliege filter topology has a fixed gain of 1. It is best to implement it from split supplies, although it can be operated from a single supply. Inject a reference into RQ instead of ground to operate off split supplies. The input and output will have to be isolated by DC blocking capacitors as with other filter types.

Many designers use the "Twin-T" notch topology of Section 6.5.4 for notches. While it is a popular topology, it has many problems. The biggest is that it is not producible. Many runs of simulation with component tolerances of 1% have shown tremendous variation in notch center frequency and notch depth. The only real advantage is that it can be implemented with a single op amp. Some additional stability can be obtained from the two op amp configuration, but if two op amps are used, then why not use a different topology such as the Fliege? To successfully use the Twin-T topology, six precision components are required. The Fliege will produce a deep null at some frequency, and it is easy to tune that frequency by adjusting one of the R_o resistors; the null will remain as deep over a fairly wide range. The response plot shown in Figure 6.13 was made by varying the potentiometer on a 10 kHz Fliege filter in 5% increments.

Some key "takeaways" from the Fliege filter response: it does not disturb the frequencies around it to any significant degree. The response in Figure 6.12 shows that a high Q 10 kHz bandpass filter leaves everything under 9 kHz and above 11 kHz almost unchanged. The response in Figure 6.13 shows that a pot forming 2% of the R_o value in the position shown allows adjustment of the center frequency over about a $\pm 1\%$ range around the center frequency. The depth of the notch

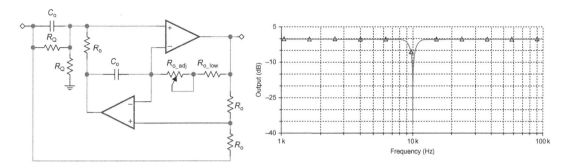

Figure 6.12
Notch Filter

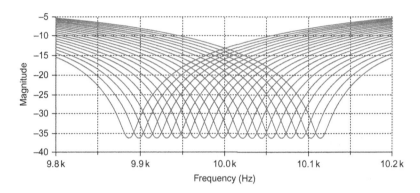

Figure 6.13
Variable-Frequency Notch Filter

remains unchanged over that range of adjustment, making it an ideal way of tuning the notch frequency.

Incidentally, if the reader wants to construct a medium-wave heterodyne filter such as the one shown above, the component values are R_o = 4.42 k, C_o = 3600 pF, R_{o_low} = 4.32 k, R_{o_adj} = 200 Ω, and R_Q = 88.7 k. The op amps should be at least 100 MHz in bandwidth.

6.4 High-Speed Filter Design

At high speeds, filter design gets particularly interesting. Of course, interesting is a word and a half word. If you love challenges, high-speed filter design will test the limits of what you can do with analog filters, with the foreknowledge that things will start to get strange!

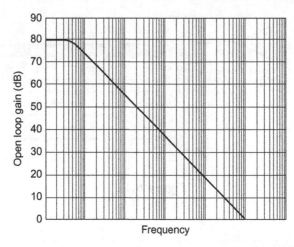

Figure 6.14
Open-loop Response

6.4.1 High-Speed Low-Pass Filters

Reviewing a figure from earlier in the book, it is obvious from Figure 6.14 that the general shape of the open-loop response plot is that of a low-pass filter. Therefore, if the low-pass breakpoint is not terribly critical, all you have to do is to select an op amp with a unity gain bandwidth at the desired -3 dB breakpoint. Imagine explaining to a manager how the low-pass filter actually is identical to a unity gain stage. The high-speed low-pass filter can also have gain, although it will lower the -3 dB breakpoint by approximately a decade for each 20 dB of gain.

6.4.2 High-Speed High-Pass Filters

I refer the reader back to a statement I made in Section 6.3.3: "There is no such thing as an active high-pass filter, but for a different reason. The gain/bandwidth product of the op amp used will ultimately produce a low-pass response characteristic, making this a wide bandpass filter. It is your responsibility to choose an op amp with a frequency limit well above the bandwidth of interest." This is doubly so at high speeds, because you are inevitably closer to the open-loop limitation of the op amp.

6.4.3 High-Speed Bandpass Filters

This is where things get really interesting, because the physics of an op amp can and will actually force the resonant peak off frequency (lower) and erode the peak.

You may be asking: how can this be? The capacitors and resistors are supposed to define the center frequency of the filter; they will not change with frequency. This is a valid question, and it leads to the answer: the frequency shift is coming from the op amp itself. But again, why? The answer comes from the location of the open-loop response characteristic of the op amp, which is the ultimate speed limit of the op amp at any frequency. A bandpass filter is composed of both low-pass and high-pass elements, and the high-pass characteristics will tend to get chopped off as they approach the open-loop characteristic. This will appear first as an amplitude limitation, and finally as a frequency shift as the bandwidth limitation of the high-pass elements comes into play and limits the point at which they interact with the low-pass filter elements. The result is a truncated response that appears to be a frequency shift.

To illustrate this effect, bandpass filters were constructed, using the topology of Section 6.3.4. The results are shown in Figure 6.15. Three frequencies were constructed, indicated by the sets of peaks in the figure. For 10 MHz, the third set of peaks, a Q (and gain) of 1, the open-loop response of the op amp was a little over 30 dB above the peak at 10 MHz, and the filter actually worked very well. As the Q (and gain in V/V) was raised in steps of 5, however, things began to change. By the time a Q of 25 was attempted, the gain of the filter was almost back to

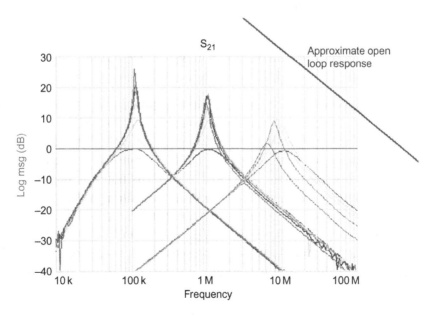

Figure 6.15
Bandpass Response

unity, and the frequency shifted to the left to about 6.5 MHz. Clearly, the proximity of the open-loop response was affecting the op amp. Even the attempts to make a 1 MHz bandpass filter — although not showing the undesirable frequency shift — still show an amplitude compression effect. Only the 100 kHz filter shows anything close to lab results matching theoretical results.

Note that the open-loop response of this particular op amp indicates that it is approximately a 1 GHz op amp — in other words, close to the state of the art in op amp design. Therefore, there is a practical limit to how fast a bandpass filter can be constructed: about 10 MHz for unity gain and Q of 1, or about 1/100 the rated bandwidth of the op amp. If a higher Q is desired, say 10, then the practical limit is about 1/1000 the rated bandwidth of the op amp. In other words:

$$\text{Center frequency (maximum)} = Aol/(100^*Q) \qquad (6.11)$$

This limitation may also hit lower frequency bandpass filters, so be extremely careful! Even a fairly low-frequency bandpass filter may require a very fast op amp to accommodate higher values of Q. You have two choices if you encounter this limit: either select an op amp with a higher gain/bandwidth product, or lower the Q of the bandpass filter.

6.4.4 High-Speed Notch Filters

A very similar bandwidth restriction affects notch filters. Instead of eroding the amplitude of the peak, as it does bandpass filters, the bandwidth restriction erodes the depth of the notch (Figure 6.16).

Using the exact same op amp as in the bandpass section above, and the notch filter topology of Section 6.3.6, notch filters were constructed at 10 MHz, 1 MHz, and 100 kHz. No center frequency tuning was attempted. The 10 MHz results were so terrible they are not included here. Even the 1 MHz results show dramatically how bandwidth was affecting the notch. At a Q of 1, a 30 dB notch is possible at 1 MHz. However, higher Q values only erode the depth of the notch further. This erosion is even evident at 100 kHz. At this point, it is clear that even a 1 GHz op amp can only be used to construct notch filters at 1 MHz and a Q of 1. If a Q of 10 is desired, a 1 GHz op amp can only be used to construct a notch filter of 100 kHz. This amazing degree of limitation was totally unexpected, to say the least.

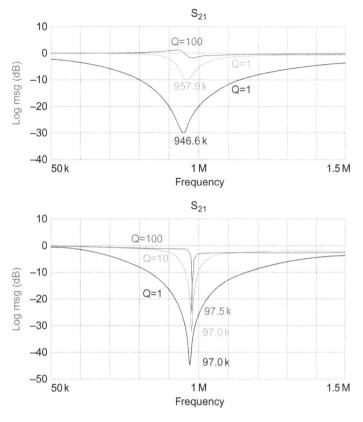

Figure 6.16
Notch Filter Response

6.5 Getting the Most Out of a Single Op Amp

As unexpected as some of the results above were, there are even stranger things that can be done with op amps. Well, some not so strange, but this chapter will not disappoint in later sections!

6.5.1 Three-Pole Low-Pass Filters

Section 6.3.2 showed how to implement low-pass filters easily and quickly. However, why implement only a two-pole filter, when a single op amp can just as easily implement a three-pole filter (Figure 6.17)?

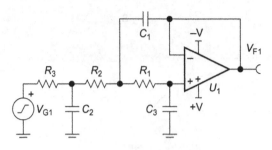

Figure 6.17
Three-Pole Low-Pass Filter

The response, in this case, will roll off 60 dB per decade instead of 40 dB per decade as seen in the two-pole filter. This topology also solves a different problem associated with the Sallen–Key architecture, that of feedthrough. In a two-pole Sallen–Key filter, high frequencies will leak through the filter, especially when the amplifier is turned off. This three-pole topology adds an RC low-pass filter to the input of a two-pole Sallen–Key architecture, thus absolutely guaranteeing at least a 20 dB per decade roll-off of high frequencies no matter what happens to the amplifier.

To design this three-pole low-pass filter:

- Pick a value of resistance $R = R_1 = R_2 = R_3$.
- Calculate a base value fsf $= 2*\pi*R*$Frequency.
- Calculate $C_1 = 3.546$/fsf.
- Calculate $C_2 = 1.392$/fsf.
- Calculate $C_3 = 0.2024$/fsf.
- Pick standard value capacitors closest to the calculated ones above.
- DONE!

6.5.2 Three-Pole High-Pass Filters

Just as it is easy to implement three-pole low-pass filters, it is also easy to implement three-pole high-pass filters (Figure 6.18):

To design this three-pole high-pass filter:

- Pick a value of capacitance $C = C_1 = C_2 = C_3$.
- Calculate a base value fsf $= 2*\pi*C*$Frequency.
- Calculate $R_1 = 3.546$/fsf.
- Calculate $R_2 = 1.392$/fsf.
- Calculate $R_3 = 0.2024$/fsf.

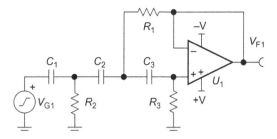

Figure 6.18
Three-Pole High-Pass Filter

- Pick standard value resistors closest to the calculated ones above.
- DONE!

6.5.3 Stagger-Tuned and Multiple-Peak Bandpass Filters

This book has purposely not covered an interesting topology that is very popular: the Twin-T topology. Part of the reason why this has not been done is that for bandpass filters, it is not a true bandpass topology. It is more of a resonator at the center frequency, and also has theoretically infinite gain at its resonance (for ideal components). Therefore, in the real world, it is difficult to control the gain at the center frequency, and the ultimate stop band rejection is 0 dB — therefore unity gain. Thus, it is not very useful in rejecting out-of-band signals.

The Twin-T topology for bandpass filters and its response are shown in Figure 6.19. This topology is very difficult to work with. It requires the user to obtain three resistors, one of which is exactly half the other, and three capacitors, one of which

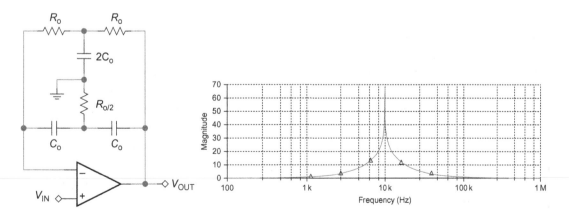

Figure 6.19
Twin-T Bandpass Filter Response

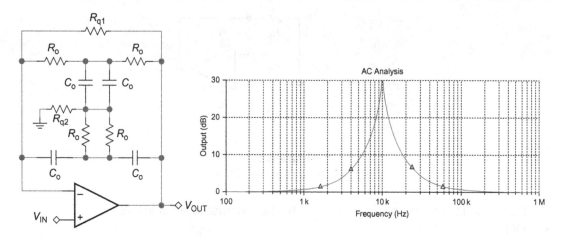

Figure 6.20
Modified Twin-T Topology

is exactly twice the other. Even if this is possible, the chances are that they would not match exactly, or track with temperature. Furthermore, the peak is so sharp that real-world components might erode the peak or miss it entirely. To overcome these shortcomings, the modifications in Figure 6.20 are made.

This configuration takes advantage of parallel resistance and capacitance characteristics to make your job easier. An additional R_o and C_o have been added, but this means that you now only have to find four identical values of resistance and four identical values of capacitance — no more 2 C_o and ½ R_o required. Because components manufactured in the same lot often have virtually identical characteristics, this should be easy.

A more subtle and unfamiliar change to the familiar Twin-T topology is the addition of R_{q1} and R_{q2}. This takes advantage of the only two places where the circuit's Q can be adjusted. As long as $R_{q1} \gg R_o$, it has minimal effect on the center frequency, but it acts to make the two C_os that are in series appear to have more leakage, therefore eroding the amplitude of the peak. Similarly, if $R_{q2} \ll R_o$, it makes the two parallel C_os appear to have higher equivalent series resistance (ESR), also eroding the peak. Acting together, it is possible to have a measure of control over the amplitude and peak, although not as precise as you might like. Comparing the response of the modified and unmodified circuits, you can see that the tendency of the amplitude to be unbounded at resonance has been mitigated, and the response is more reasonable. The Q has also been modified, although it is hard to see on the 4 decade log frequency scale.

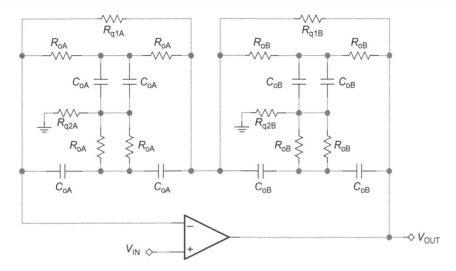

Figure 6.21
Two Twin-T Networks Inside the Feedback Loop

Another technique that has been used to adjust the Q of a Twin-T circuit is to unbalance R_o and C_o in the legs of the circuit that are in series. This, however, requires you to find multiple precision component values, all of which could potentially track differently over temperature.

With this brief introduction to Twin-T bandpass filters complete, now the real fun can begin! There is no reason that two Twin-T networks cannot be placed inside the feedback loop of an op amp (Figure 6.21). The two Twin-T sections are effectively in series inside the feedback loop of the op amp. They can be tuned entirely independently, resulting in the response shown in Figure 6.22.

Such a strategy is often employed when it is necessary to have flat phase (and group delay) in a narrow region around the center frequency. The trade-off is that the passband of the filter is widened to accommodate the flat phase response in the middle. You are also reminded to select a wide band op amp that will be able to handle not only the gain at the center frequency, but also the gain of the upper peak. The open-loop response should be at least 40 dB above the upper peak. In this case, an op amp with 1.5 GHz bandwidth was used to simulate a stagger tuned 10 kHz filter, in other words a bandwidth more than five orders of magnitude greater than the center frequency.

If the two Twin-T networks are separated enough in frequency, this topology will produce a bandpass filter that has more than one resonant peak, as shown in Figure 6.23. Notice that the flat phase response has been sacrificed between the

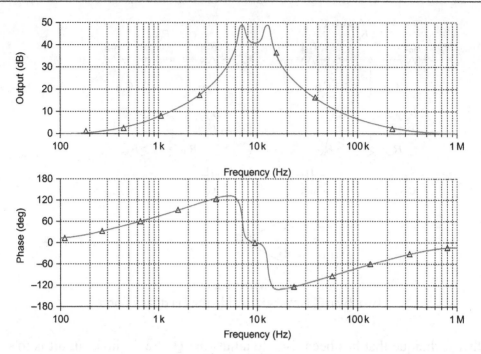

Figure 6.22
Stagger-Tuned Filter Response

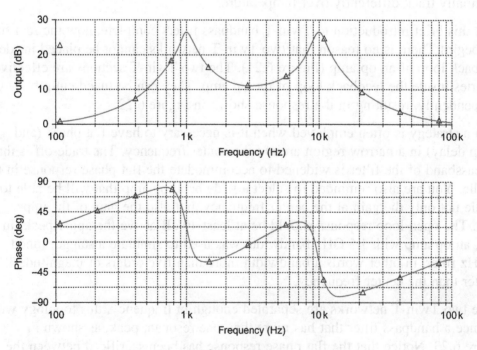

Figure 6.23
Multiple-Peak Bandpass Filter

peaks, although the phase still crosses through 0 degrees. But a flat phase between the peaks is no longer the design goal; rather, having two peaks is. The "valley" between the two peaks is not that low; it would be lower if the peaks were further separated in frequency or the Q was higher. But for quick detection of multiple tone frequencies, this method is the most economical possible.

You are cautioned again about op amp bandwidth, but in this case the open-loop gain at the upper peak may be less of a constraint because the overall gain is lower.

6.5.4 Single-Amplifier Notch and Multiple-Notch Filters

The Twin-T topology can also be used to create a single amplifier notch filter (Figure 6.24). As was the case in the Twin-T bandpass filter, two resistors were added that allow some control over the Q, also affecting notch depth. This is very useful, because the Twin-T notch is difficult to tune for center frequency. If the Q is lowered, the chances are better of actually placing a notch where it is needed to reject an unwanted frequency.

Unfortunately, the Twin-T topology also has the drawback that the notch disturbs the amplitude in the decades above and below the center frequency. Contrast the response above with the response in Figure 6.24. Where the Fliege topology leaves the surrounding frequencies almost untouched, the Twin-T has a very significant effect almost a decade above and below the center frequency, making it a poor choice for a notch filter. So, why use it at all? Besides the advantage of being a single amplifier notch, the other advantage is, like the Twin-T bandpass topology,

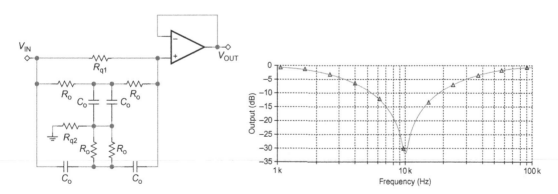

Figure 6.24
Single-Amplifier Twin-T Notch Filter

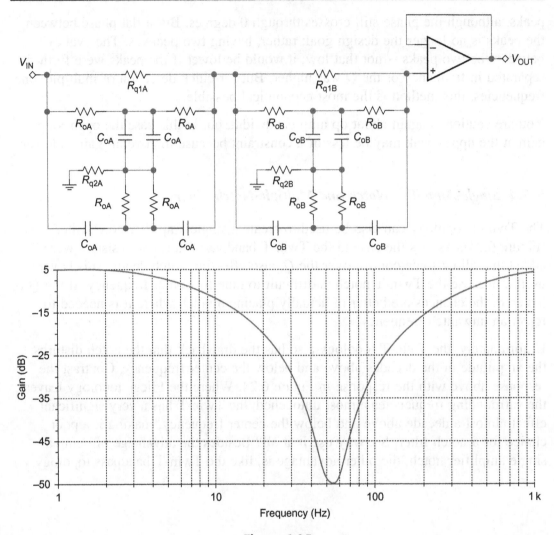

Figure 6.25
Twin-T Band Reject Filter

that it is possible to produce two notches from a single amplifier. This can be a way either to produce two individual notches at widely separated frequencies, or to produce a band rejection filter if the notch frequencies are closely spaced (Figure 6.25).

This scheme dramatically affects the audio above 10 Hz and below 300 Hz. It may, however, be suitable for severe cases of hum on speech.

6.5.5 Combination Bandpass and Notch Filters

There is no reason why a Twin-T notch and a bandpass filter cannot be combined in a single circuit (Figure 6.26).

Of course, more than one peak and/or notch is also possible. In practice, however, you should probably count on no more than two or three sections of Twin-T networks, because the number of passive components crowding the op amp will become excessive, leading to parasitics on the board and other problems.

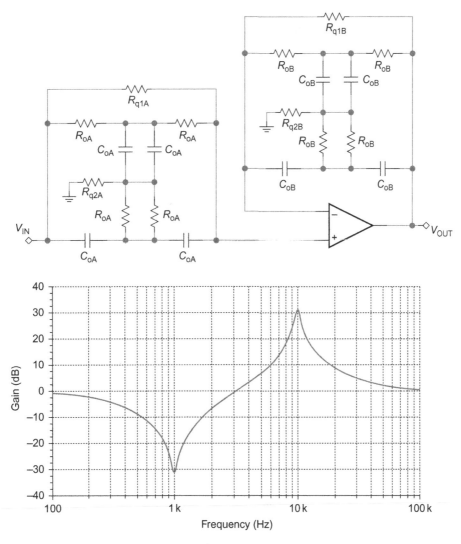

Figure 6.26
Single-Amplifier Twin-T Notch and Bandpass Filter

6.6 Biquad Filters

No three- and four-amplifier biquad circuits are presented here. The overwhelming majority of applications require you to use as few op amps as possible to save power, board space, and cost; reduce noise; and improve reliability. These advantages usually far outweigh any advantages a multiple op amp configuration might provide, such as providing more than one output function at one time, or allowing separate, independent adjustments of each filter parameter. If you need such a filter, there are many volumes out there covering multiple amplifier biquads. I will offer only one small piece of advice here: consider using a fully differential amplifier in biquad configurations. Often the third amplifier in a biquad circuit serves only to invert the signal; fully differential op amps already include that inverted output.

6.7 Design Aids

Just as I did for op amp gain and offset, I have written design aids for filter design.

6.7.1 Low-Pass, High-Pass, and Bandpass Filter Design Aids

This chapter has presented you with a bewildering variety of filter circuits to ponder. It is impossible (well, it would be very difficult) to produce a single circuit that could implement all filter types. I will leave you instead with a single schematic of a universal filter circuit (well, low pass, high pass, and band pass) (Figure 6.27).

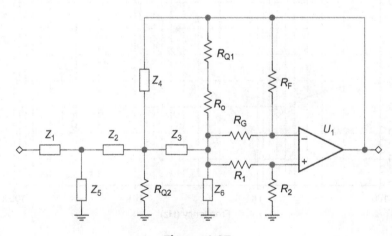

Figure 6.27
Universal Filter Schematic

The general-purpose impedances, or Z, can be either resistors or capacitors, depending on the filter topology selected. Not all components will be installed for every filter. Some may be 0Ω resistors, whereas others may be "open", not installed.

This single circuit can implement low-pass, high-pass, and bandpass stages. Although two-pole Sallen—Key filters can be implemented, why bother when you can have three poles just as easily?

As was the case for gain and offset, I have written a design aid for filters, which is also available on the companion website. Figure 6.28 shows a screen shot. Some explanation is in order. When the calculator first comes up, all of the impedances and units will be blank, unlike the screen shot in the figure. To use the calculator, first you need to know whether you are designing a low-pass, high-pass, or bandpass filter. These three types are available in a drop-down menu in the "Filter Type" box. Then, there are a series of boxes that depend on the type of filter you are making: For (type of filter), Enter (value) series of lines. For example, all filter types, low-pass, high-pass, and bandpass, require a frequency. For low and high pass, it will be the −3 dB cut-off frequency. For band pass, it will be the center frequency. So, for ALL filters, Enter Frequency. Because this is a low-pass filter, the only remaining decisions you need to make

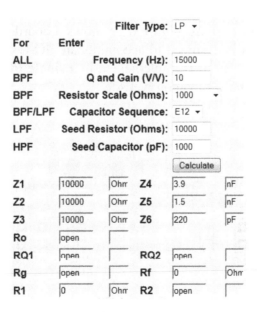

Figure 6.28
Universal Filter Calculator

are the capacitor sequence (E6, E12, or E24) and the Seed Resistor value, which will scale the capacitors. Click the "Calculate" button and the component values appear.

If you choose a high-pass filter instead, you will notice that capacitors and resistors have swapped left to right. Of course, you choose a seed capacitor, instead of a seed resistor, for high-pass filters. The seven resistor values on the bottom, $R_o - R_2$, are the same for low-pass and high-pass filters, either 0Ω or open. These route connections configure the universal schematic and PCB to be a three-pole Sallen−Key filter topology. This changes dramatically when you choose "band pass" instead. Z_1 becomes 0Ω, Z_5 and Z_6 become open, and some of the resistor values on the bottom are populated, while others are used to configure the circuit for the modified Deliyannis configuration.

In addition, the schematic and circuit notes will appear to the right of the calculator. These change according to topology. It will take some experience on your part to properly scale resistors and capacitors. All capacitors should be 1% NPO/COG dielectric if at all possible, especially if the filter is to encounter different temperatures. Although higher value NPO/COG capacitors are becoming more common, 10,000 pF is generally the limit, and anything above 1000 pF is usually large, expensive, and hard to obtain. Resist at all costs the temptation to use 5% capacitors, especially in the bandpass filter. Choose 1% tolerance, and make that decision stick with procurement and manufacturing departments.

To further facilitate your designs, Figure 6.29 shows a board layout of the universal filter schematic above. It is designed to accommodate SO-8 single op amps and 1206 surface-mount resistors. It is implemented in a single PCB layer. The gerbers are also available from the companion website.

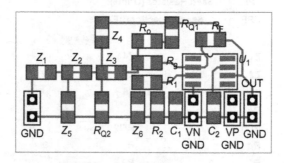

Figure 6.29
Universal Filter Board

You can use this approach to implement a universal arrangement on a PCB. This might be done when multiple filter stages are to be used, and it is unclear whether a given digital signal processor (DSP) algorithm really needs a high-pass function or a low-pass function, or whether the application can benefit from two stages of one type, for example.

6.7.2 Notch Filter Design Aids

But wait, there is more! Remember I said it would be difficult to implement a universal filter board that also contained a notch? I have not left you to your own devices. I also have written a notch filter design aid. A screen capture is shown in Figure 6.30.

Like the other filter calculator, this one also shows the schematic. This calculator only requires a center frequency; the user can adjust the Q, capacitor sequence, and resistor scale. As the notes indicate, however, you should not deviate too much from a Q of 10: this is a good compromise value that will give plenty of rejection, yet leave other frequencies untouched. The calculator also includes provisions for tuning the center frequency. If you do not want to do that, just substitute another R_o for R_{o_low} and R_{o_adj}.

I have also generated a PCB layout for the notch filter (Figure 6.31). This one had to be two layers, mainly so I could route the power. If this is a hardship, just route the power with wires on the back of a single-layer board.

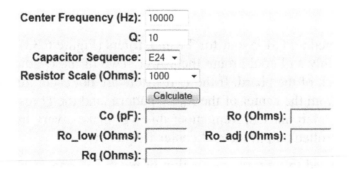

Figure 6.30
Notch Filter Calculator

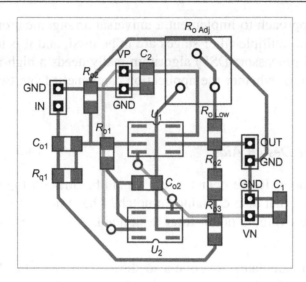

Figure 6.31
Notch Filter PCB

6.7.3 Twin-T Design Aids

I have written separate design aids for Twin-T bandpass and Twin-T notch filters. The screen capture for both of them looks the same (Figure 6.32).

There is little to decide: enter the center frequency, select the resistor and capacitor sequence, and click "Calculate". There is a resistor scale control. Leave the Q alone; it will allow about 30 dB gain for the band pass and 30 dB of rejection for the notch. Adjustment is possible, but over a relatively narrow range. Remember that Twin-T filters cannot be tuned easily for center frequency, and the response will not be ideal.

I have also generated a PCB layout for Twin-T filters (Figure 6.33). This one had to be two layers, mainly so I could route the power. Two of the Q adjustment resistors are also on the back of the board. If the Q resistors are not used, jumpers can be added to ground from the center of the R_o C_o pattern, and the Q resistors on the top of the board can be left unused. Remember this will make a very narrow bandpass/notch with uncontrolled amplitude and center frequency.

The board is designed to support one section of notch and one section of band pass. If notch is desired, place a 0Ω resistor across the R_{q3} pads. If band pass is desired, place a 0Ω resistor across the R_{q1} pads. This board will also support the notch/bandpass configuration of Section 6.5.5.

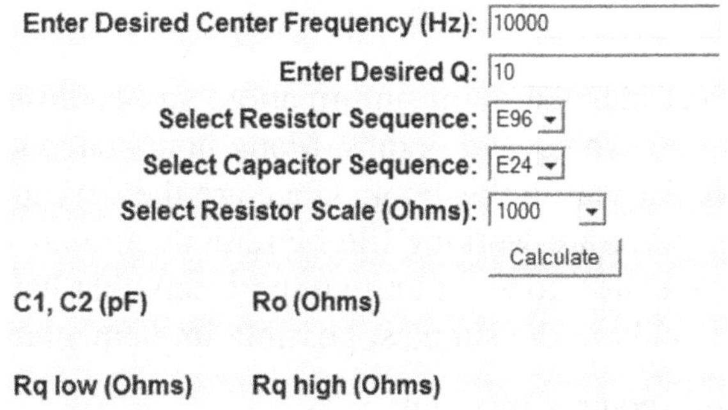

Enter Desired Center Frequency (Hz): 10000
Enter Desired Q: 10
Select Resistor Sequence: E96
Select Capacitor Sequence: E24
Select Resistor Scale (Ohms): 1000
Calculate
C1, C2 (pF)
Ro (Ohms)
Rq low (Ohms)
Rq high (Ohms)

Figure 6.32
Twin-T Filter Calculator

R_{q3}
C_{o7}
R_{o6}
R_{o7}
C_{o8}
R_{q1}
R_{o5}
C_{o5}
C_{o6}
R_{o8}
C_{o1}
R_{o1}
R_{o2}
C_{o2}
IN
GND
OUT
GND
C_1
VN
VP
C_2
R_{o3}
C_{o3}
C_{o4}
R_{o4}
GND GND

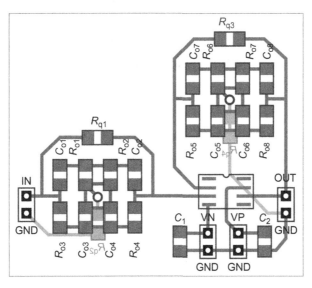

Figure 6.33
Twin-T PCB

6.7.4 Final Comments on Filter Design Aids

You can quickly implement an entire signal chain using multiples of these filter boards and the universal gain and offset board of Chapter 3. Care must be taken, of course, to get the DC operating point of the circuit right, so you might want to implement filters first, then gain and offset.

6.8 Summary

Filter design is a subject plagued by misinformation, and endless textbooks with mathematical derivations, tables, and graphs. Many hours have gone into demystifying the topic for you in this book. Gone are the derivations and math exercises, and in their place are working filter circuits with real-world component values. Many hours have also gone into providing filters that have been tested extensively in the lab, and are the simplest possible implementation available.

Filter design in general requires that you:

- know which frequencies you want to pass, and which frequencies you want to reject
- have the ability to choose a filter topology that accomplishes the passing and rejection of those frequencies
- have the ability to calculate the component values to accomplish that filtering function − it is hoped that the design aids will greatly facilitate this.

When this procedure is followed, good results follow. As you design filters successfully, you will gain confidence and experience in the complex subject of filter design.

CHAPTER 7

Using Op Amps for Radio frequency Design

7.1 Introduction

Radio frequency (RF) design used to be the exclusive domain of discrete devices. The advent of new generations of high-speed voltage and current feedback op amps has made it possible to use op amps for RF design. Op amp-based RF circuitry is easier to design, and has less risk associated. "Tweaking" in the lab can be almost eliminated. Although there are many advantages, traditional RF designers are reluctant to utilize them. They are confronted with a bewildering array of op amp parameters, many of which do not relate directly to the set of design parameters with which they are familiar. This chapter bridges the gap between RF designers and op amp designers, giving the RF designer common ground with which to begin their design.

7.2 Voltage Feedback or Current Feedback?

The RF designer considering op amps is presented with a dilemma: are voltage feedback amplifiers or current feedback amplifiers better for the design? Frequency of operation is usually the most demanding aspect of RF design, and this makes the op amp bandwidth a critical parameter. The bandwidth specification given in op amp data sheets only refers to the point where the unity gain bandwidth of the device has been reduced by 3 dB by internal compensation and/or parasitics, which is not very useful for determining the actual operating frequency range of the device in an RF application.

Internally compensated voltage feedback amplifier bandwidth is dominated by an internal "dominant pole" compensation capacitor. This causes a constant gain/bandwidth limitation. Current feedback amplifiers, in contrast, have no dominant pole capacitor, and therefore can operate much closer to their maximum frequency at higher gain. Stated another way, the gain/bandwidth dependence has been broken, but not to the degree most designers would want. In practice, current feedback op amps offer only a slight advantage over voltage feedback op amps.

7.3 Radio frequency Amplifier Topology

The traditional RF amplifier shown in Figure 7.1 uses a transistor (or in the early days a tube) as the gain element. DC bias ($+ V_{bb}$) is injected into the gain element at the load through a bias resistor R_b. RF is blocked from being shorted to the supply by an inductor L_c, and DC is blocked from the load by a coupling capacitor.

Both the input impedance and the load are 50Ω, which ensures matching between stages.

When an op amp is substituted as the active circuit element, several changes are made to accommodate it.

By themselves, op amps are differential input, open-loop devices. They are intended to be operated in a closed-loop topology, different from a receiver's automatic gain control (AGC) loop. The feedback loop for each op amp must be closed locally, within the individual RF stage.

There are two ways to close an op amp locally: inverting and non-inverting. These terms refer to whether the output of the op amp circuit is inverted from the input or

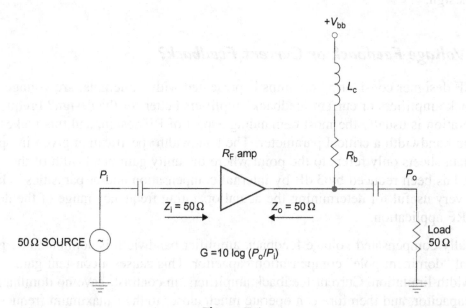

Figure 7.1
A Traditional Radio frequency Stage

not. From the standpoint of RF design, this is seldom of any concern. For all practical purposes, either configuration will work and give equivalent results, but RF designers may find the non-inverting configuration of Figure 7.2 easier to work with, because the gain setting elements are not part of impedance matching.

The input impedance of the non-inverting input is high, so the input is terminated with a 50Ω resistor. Gain is set by the ratio of the feedback resistor (R_f) and the gain resistor (R_g):

$$G = 20 \cdot log \frac{1}{2}\left(1 + \frac{R_f}{R_g}\right) dB, \log \text{ gain} \tag{7.1}$$

For a desired gain:

$$1 + \frac{R_f}{R_g} = 2\left(10^{G/20}\right) \tag{7.2}$$

The gain of this stage as shown should never be below ½ (-6 dB), because most op amps are unity gain stable.

The output of the stage is converted to 50Ω by placing a 50Ω resistor in series with the output. This, combined with a 50Ω load, means that the gain is divided by 2

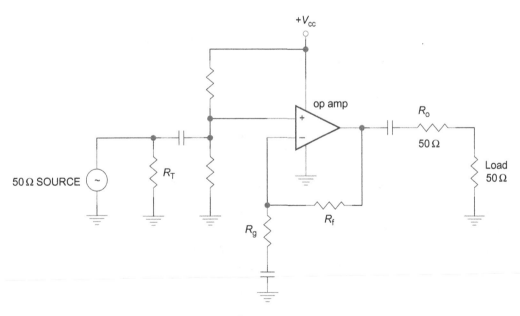

Figure 7.2
Non-Inverting Radio frequency Op Amp Gain Stage

(-6 dB) in a voltage divider. So a unity gain (0 dB) gain stage would become a gain of ½, or -6 dB.

A virtual ground is generated on the non-inverting input after the coupling capacitor, to raise the operating point of the op amp to a virtual ground halfway between the supply voltage and ground.

Coupling capacitors are needed to isolate preceding and succeeding stages, and the gain resistor R_g from the virtual ground DC potential. These capacitors should be selected to have low impedance at the operating frequency, but not so small that they affect the gain of the stage directly, or cause unacceptable variations in gain over the operating range of the stage.

7.4 Op Amp Parameters for Radio frequency Designers

When designing for RF, problems can arise because op amp sheets show a totally different set of parameters than are considered for RF designs. This situation has been improving over the years as more and more high-speed amplifiers have been introduced, but there are still differences. This section will go over the more important parameters and specifications, and show RF designers how to interpret op amp data sheet specifications to meet their needs.

7.4.1 Stage Gain

Op amp designers think of the gain of an op amp stage in terms of voltage gain. RF designers, in contrast, are used to thinking of RF stage gain in terms of power:

$$\text{Absolute power } (W) = \frac{V_{RMS}^2}{50\ \Omega}$$

$$P_o(\text{dBm}) = 10 \cdot log\left(\frac{\text{Absolute power}}{0.001\ W}\right) \tag{7.3}$$

$$\text{dBm} = \text{dBV} + 13 \text{ in a 50 } \Omega \text{ system}$$

The forward transmission S_{21} is specified over the operating frequency range of interest. S_{21} is never specified on an op amp data sheet because it is a function of the gain, which is set by the input and feedback resistors R_f and R_g. The forward transmission of a non-inverting op amp stage is:

$$S_{21} = A_L = \frac{V_L}{V_i} = \frac{1}{2}\left(1 + \frac{R_f}{R_g}\right) \tag{7.4}$$

Op amp data sheets show open-loop gain and phase. It is the responsibility of the designer to know the closed-loop gain and phase. Fortunately, this is not hard to do. The data sheets many times include excellent graphs of open-loop bandwidth, and most of the time include phase. Closing the loop produces a straight line across the graph at the desired gain, curving to meet the limit. The open-loop bandwidth plot should be used as an absolute maximum.

An added benefit of using current feedback amplifiers is that the values of R_f and R_g are specified in the data sheet. Please note, however, that the gains in the data sheet will not take 50Ω matching into account, so the stage gains will be half what the data sheet specifies for a given value of R_f and R_g.

7.4.2 Phase Linearity

Often, a designer is concerned with the phase response of an RF circuit. This is particularly the case with video design, which is a specialized type of RF design. Current feedback amplifiers tend to have better phase linearity than voltage feedback amplifiers:

- voltage feedback THS 4001: differential phase $= 0.15°$
- current feedback THS 3001: differential phase $= 0.02°$.

7.4.3 Frequency Response Peaking

Current feedback amplifiers allow an easy resistive trim for frequency peaking that has no impact on the forward gain. Figure 7.3 shows this adjustment added to a non-inverting circuit. This resistive trim inside the feedback loop has the effect of adjusting the loop gain, and hence the frequency response without adjusting the signal gain, which is still set by R_f and R_g.

Values for R_f and R_g must be reduced to compensate for the addition of the trim potentiometer, although their ratio and hence the gain should remain the same. The adjustment range of the pot, combined with the lower R_f value, ensures that the frequency response can be peaked for slight current feedback amplifier parameter variations.

7.4.4 − 1 dB Compression Point

The -1 dB compression point is defined as the output power, at a fixed input frequency, where the amplifier's actual output power is 1 dBm less than expected.

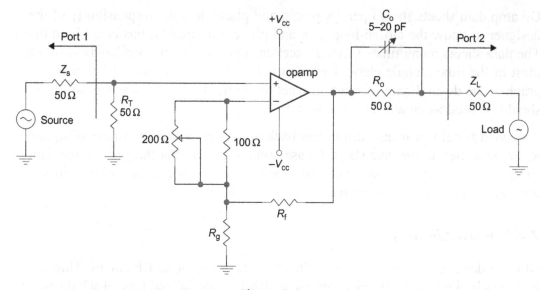

Figure 7.3
Frequency Response Peaking

Stated another way, it is the output power at which the actual amplifier gain has been reduced by 1 dB from its value at lower output powers. The -1 dB compression point is the way RF designers talk about voltage rails.

Op amp designers and RF designers have very different ways of thinking about voltage rails, which are related to the requirements of the systems that they design:

- An op amp designer, interfacing op amps to data converters, for example, takes great pains not to hit the voltage rail of the op amp, thus losing precious codes.
- An RF designer, on the other hand, is often concerned with squeezing the last half decibel out of an RF circuit. In broadcasting, for example, a very slight increase in decibels means a lot more coverage. More coverage means more audience and more advertising dollars. Therefore, slight clipping is acceptable, as long as resulting spurs are within US Federal Communications Commission (FCC) regulations.

Standard AC-coupled RF amplifiers show a relatively constant -1 dB compression power over their operating frequency range. For an operational amplifier, the maximum output power depends strongly on the input frequency. The two op amp specifications that serve a similar purpose to -1 dB compression are V_{OM} and slew rate.

At low frequencies, increasing the power of a fixed frequency input will eventually drive the output "into the rails": the V_{OM} specification. Often, V_{OM} is broken up into separate low and high clipping levels, which are specified as V_{OL} and V_{OH}. At high frequencies, op amps will reach a limit on how fast the output can transition (respond to a step input). This is the slew rate limitation of the amplifier. The op amp slew rate specification is divided by two, because of the matching resistor used at the output.

As is the case for op amps used in any other application, it is probably best to avoid operation near the rails, as the inevitable distortion will produce harmonics in the RF signal — harmonics that are probably undesirable for FCC testing. That said, if harmonics are still at an acceptable level at the -1 dB compression point, it can be a very useful way to boost power to a maximum level out of the circuit.

7.4.5 Noise Figure

The RF noise figure is the same thing as op amp noise, when an op amp is the active element. There is some effect from thermal noise in resistors used in RF systems, but the resistor values in RF systems are usually so small that their noise can be ignored.

Noise for an op amp RF circuit is dependent on:

- the bandwidth being amplified
- gain.

This example assumes the $11.5\,\text{nV}/\sqrt{\text{Hz}}$ op amp. The application is a 10.7 MHz intermediate-frequency (IF) amplifier. The signal level is 0 dBV. The gain is unity.

Figure 7.4 is extrapolated from real data. The $1/f$ corner frequency, in this case, is much lower than the bandwidth of interest. Therefore, the $1/f$ noise can be completely discounted (assuming that filtering removes any noise that would cause the amplifier or data converter to saturate).

For narrow bandwidths, noise may be quite low. Various bandwidths are shown in Table 7.1. There is an advantage to be had from reducing the bandwidth. Noise is amplified by the gain of the stage. Therefore, if a stage has high gain, care must be taken to find a low-noise op amp. If the gain of a stage is lower, then the noise will not be amplified as much, and a less expensive op amp may be suitable.

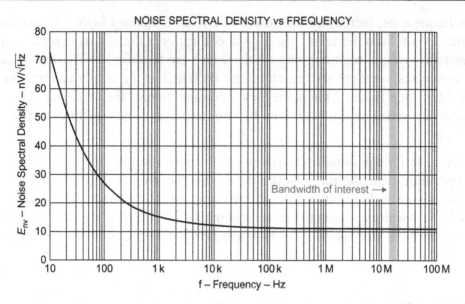

Figure 7.4
Noise Bandwidth

Table 7.1: Noise for Various Bandwidths

Bandwidth (kHz)	E_{in} (µV)	S/N (dB)
280	6.09	−104.3
230	5.52	−105.2
180	4.88	−106.2
150	4.45	−107.0
110	3.81	−108.4
90	3.45	−109.2

7.5 Wireless Systems

Figure 7.5 shows an example of a dual-IF cellular receiver. The receiver contains two mixer stages reminiscent of a classic double conversion superheterodyne receiver.

The RF designer wishing to use op amps for some of the gain stages needs to realize the current state of the art in op amp technology. Op amps are excellent choices for driving an analog-to-digital converter (ADC), low-pass filtering, and even second IF gain. But when you attempt to use op amps for the first IF frequency of 660 MHz in Figure 7.5, you will quickly run out of open-loop gain. This limits the RF designer to using op amps in the final stages of that design,

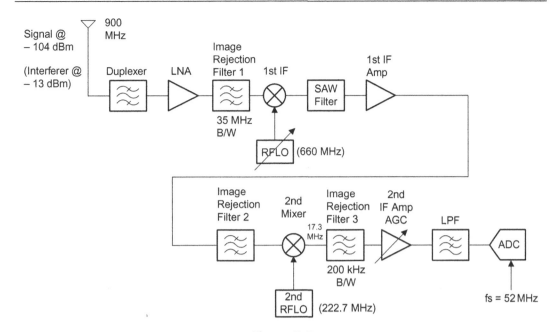

Figure 7.5

Typical Global System for Mobile Communications (GSM) Cellular Base Station Receiver Block Diagram

although unity gain buffers can be employed for impedance matching at higher frequencies. For the most part, though, op amps are not suitable for ultra-high-frequency (UHF) applications.

7.5.1 Broadband Amplifiers

Current feedback amplifiers are the component of choice for broadband amplifiers. The THS3202 was tested for its wide bandwidth and fast slew rate. A THS3202 evaluation board made a convenient platform on which to construct the circuits presented here. The key questions explored just how much gain an op amp-based circuit is capable of, and over what frequency range. The circuit in Figure 7.6 was used to explore these questions. It was first configured as a single stage, consisting of device A of a dual package, a 301Ω feedback resistor, a 16.5Ω gain resistor, and a 49.9Ω back-termination resistor. This configuration produces an amplifier gain of 20, a stage gain of 10 when connected to a 50Ω monitoring device.

Note the simplicity of this circuit compared to traditional RF circuitry. Provide the op amp, termination and decoupling components, and two resistors, and the circuit is complete. The 301Ω (R_f) and 16.5Ω (R_g) resistors are all that are required to set the stage gain.

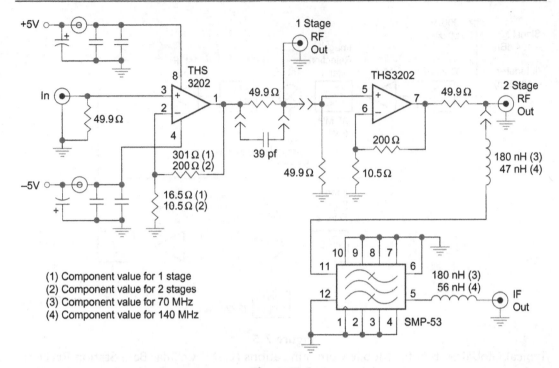

Figure 7.6

Broadband Radio frequency Intermediate-Frequency Amplifier

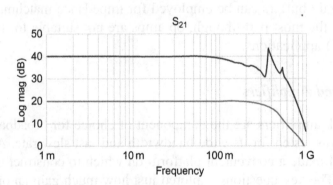

Figure 7.7

Wideband Response

The stage's gain can be set precisely by the resistors alone, and this is one of the op amp-based design's strong points. This circuit produces the lower amplitude curve in Figure 7.7.

The op amp stage's voltage gain itself is 20, but this is cut in half by the back-termination resistor's action in combination with the load. The RF amplifier's

−3 dB point is about 390 MHz. If a flat gain over frequency is required, this circuit is only usable to about 200 MHz. Input and output voltage standing wave radio (VSWR) values are better than 1.01:1 for most of the bandwidth, only degrading to about 1.1:1 near 200 MHz. S_{12} is −75 dB over most bandwidth, only degrading to −50 dB near the bandwidth limit.

One might wonder whether more gain could be coaxed from the stage by lowering the gain resistor (R_g) even more. The answer is yes, but there is a practical limit. Remember that the feedback resistor (R_f) is a large determining factor for current feedback amplifier stability. Remember also that R_g has to drop proportionally more. One can see that it would not be long until the value of R_g value becomes impractically small. Lab tests were attempted with various R_f values. The results indicated there is no advantage to making it smaller than 200Ω. Below that, peaking starts to occur regardless of the value of R_g, becoming worse and worse as the resistance is made lower and lower. This is exactly what one would expect, because one of the things a designer cannot do in working with current feedback amplifiers is to make R_f a short.

More gain requires cascading multiple stages of THS3202 op amps. Fortunately for the designer, the THS3202 is a dual device, making a two-stage RF amplifier easy to implement at very little additional cost.

To convert the amplifier to two stages, the feedback resistor is lowered to 200Ω, and the gain resistor lowered to 10.5Ω. A second stage is then connected to the first, using the same values for the feedback and gain resistors. Isolation is accomplished by using interstage termination resistors. The optional 39 pF capacitor provides peaking to compensate for some high-frequency roll-off. Unfortunately, it also creates capacitative load on the first amplifier output. This increases the first stage's tendency to peak, as seen in the upper curve of 0. This indicates a tendency towards instability, and it manifests itself by poorer IP3 performance. If maximum IP3 performance is needed, the designer should delete the capacitor and live with less bandwidth from the stage. The other S parameters for this circuit are similar to those applying to the single op amp's case.

At this point, it is important to talk about signal levels. All of the above points in the direction of impressive performance, but if these benefits only apply to very small signals they are hardly benefits at all!

The signal level a designer can pass through an op amp is determined by its input and output voltage rails, as described by the device's data sheet. They form a set of high- and low-voltage "hard clipping" points for the signal as it passes through the operational amplifier. Consequently, the −1 dB compression point occurs very soon after any voltage rail limitation has been breached. The wise designer will not attempt to squeeze that last decibel from the stage because the hard clipping points may produce substantial harmonics.

For a THS3202 stage, the amplifier's output can swing ± 3.2 V; therefore the output of the stage can swing ± 1.6 V. This corresponds to 14 dBm of output power.

7.5.2 Intermediate-Frequency Amplifiers

The gain circuit shown in Figure 7.6 can be cascaded with surface acoustic wave (SAW) filters to form high-performance IF stages. The only design consideration is the insertion loss of the filter, which may not be a constant value from part to part, or from batch of parts to batch of parts. If precise gain is needed from the stage, the designer may need to include a trim resistor in one or both stages. This trim adjustment, however, will not affect the stage's tuning except for a slight effect on its upper frequency limit.

It is easy to cascade the dual wideband RF amplifier with SAW filters. Sawtek conveniently provided an EVM board, which greatly simplified prototyping. All that was required was to provide a short SMA-to-SMA cable between the two boards. It is important to place the SAW filter element after the gain stage so noise generated in the op amp circuitry is filtered with the same response as the signal. If the gain stage were placed after the SAW filter, the amplifier stage's broadband noise response would be passed to the next stage instead of a filtered, narrow-band response.

The 70 MHz and 140 MHz IF amplifiers find use in cellular telephone base stations and satellite communications receivers. A Sawtek 854660 filter was selected for 70 MHz, and a Sawtek 854916 filter for 140 MHz. These filters require input and output inductors, and operate with standard 50Ω input and output. A 70 MHz SAW filter produces the upper response curve and the 140 MHz filter produces the lower response curve shown in Figure 7.8.

The most outstanding feature of the response curves featured in Figure 7.8 is that they are almost identical in shape to the filter curves provided by Sawtek. Narrow band response is shown in 0, but there is virtually no harmonic content when the broadband response is examined. In other words, the amplifier is providing gain while not adding undesirable harmonic content. The insertion loss of the 70 MHz SAW filter is about 7 dB. The insertion loss of the 140 MHz SAW filter is only 8 dB, but the gain circuit itself is starting to roll off at this frequency, accounting for the rest of the loss. Careful examination of the lower passband shows the slight roll-off due to the broadband stage characteristic.

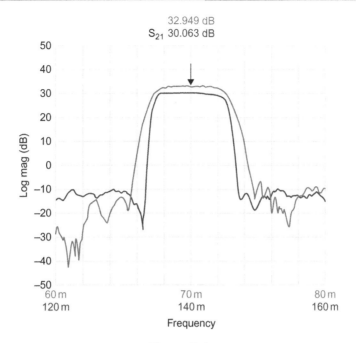

Figure 7.8
Intermediate-Frequency Amplifier Response

7.6 High-Speed Analog Input Drive Circuits

Communication ADCs, for the most part, have differential inputs and require differential input signals to properly drive the device. Drive circuits are implemented with either RF transformers or high-speed differential amplifiers with large bandwidth, fast settling time, low output impedance, good output drive capabilities, and a slew rate of the order of 1500 V/μS. The differential amplifier is usually configured for a gain of 1 or 2 and is used primarily for buffering and converting the single-ended incoming analog signal to differential outputs. Unwanted common-mode signals, such as hum, noise, DC, and harmonic voltages, are generally attenuated or cancelled out. Gain is restricted to wanted differential signals, which is often 1–2 V.

The analog input drive circuit, as shown in Figure 7.9, employs a THS4141 device. This device offers fast speed, linear operation over a wide frequency range, and wide power-supply voltage range, but draws slightly more current than a BiCMOS device. The −3 dB bandwidth is 120 MHz measured at the output of the amplifier. The analog input V_{IN} is AC-coupled to the THS4141 and the DC voltage V_{OCM} is the applied input common mode voltage. The combinations R47−C57 and R26−C34 are selected to meet the desired frequency roll-off. If the input signal

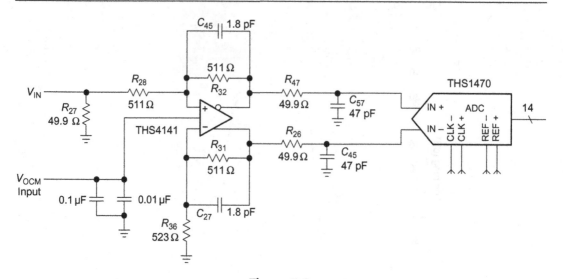

Figure 7.9
Single-Ended to Differential Output Drive Circuit

frequency is above 5 MHz, higher order low-pass filtering techniques (third order or greater) are employed to reduce the op amp's inherent second harmonic distortion component.

7.7 Conclusions

Op amps are suitable for RF design, provided that the cost can be justified by the flexibility they offer the designer. They are more flexible to use than discrete transistors, because the biasing of the op amp is independent of the gain and termination. Current feedback amplifiers are more suitable for high-frequency, high-gain RF design, because they do not have the gain/bandwidth limitation of voltage feedback op amps.

Scattering parameters for RF amplifiers constructed with op amps are very good. Input and output VSWR are good because the effects of termination and matching resistors can be made independently of stage biasing. Reverse isolation is very good, because the RF stage is made of an op amp consisting of dozens or hundreds of transistors, instead of a single transistor. Forward gain is very good with a current feedback amplifier.

Special considerations apply to RF design that do not normally apply to op amp design: the phase linearity, the −1 dB compression point (as opposed to voltage rails), the 2 tone, third order intermodal intercept, peaking, and noise bandwidth. In just about every case, the performance of an RF stage implemented with op amps is better than that of one implemented with a single transistor.

CHAPTER 8

Designing Low-Voltage Op Amp Circuits

8.1 Introduction

Op amp application circuits are a means to an end — the end being a product that is sold. More and more products are being manufactured that operate off batteries. Consumers want smaller, lighter products that still have the functionality of previous generations of products, the cell phone being an excellent example. The four-pound brick-shaped cell phone with an antenna has been supplanted by ever smaller "smart" phones. Cell phones may not be the best example because op amps are integrated inside integrated circuits (ICs) that are highly integrated. Make no mistake, though, op amp gain and filter circuits still exist inside those ICs. At some point, a designer still had to design gain and filter circuits that were eventually integrated into an IC. Since these ICs operate off a single battery, they are almost certainly single-supply circuits, harking back to Chapter 3.

The trend is towards not only single supplies, but also lower supply voltages. The first Regency transistor radio operated off a 22.5 V battery, my childhood transistor radio operated off a 9 V battery, and my latest portable operates off a single 1.5 V battery, and yet has longer battery life, and is smaller, lighter, and more sensitive than my previous radio. Thus a secondary trend is towards lower power consumption so a product operates longer between battery changes or battery charges, yet with no compromise in performance.

This trend has meant lower voltage rails and power consumption inside op amps. The familiar ± 15 V supplies have been augmented with product offerings that operate off lower and lower supply voltages, and are often not even capable of operating off ± 15 V. At first glance, you might think that this is great: no more requirement for separate analog supplies; it is wonderful to be able to run op amps off the same voltage as logic circuits. But there is a downside to operating off a lower supply voltage. This book has already introduced you to the op amp output voltage rail specifications, V_{OH} and V_{OL}. These become more and more critical as op amp supply voltages go lower and lower. There are similar limitations on the input voltage range. This chapter will give you the tools you need to make design decisions that will use the voltage ranges available, and optimize those designs to work well within a power supply "budget".

8.2 Critical Specifications

There are several specifications that are critical to understand when working on systems that have a limited power supply range. These will be discussed in order of importance.

8.2.1 Output Voltage Swing

Rail-to-rail output (RRO) voltage swing is desirable for at least two reasons. First, the dynamic range can achieve the maximum obtainable value if the op amp is RRO. Second, RRO op amps can drive any converter connected to the same power supply if the impedance is compatible. The schematic of an RRO op amp output stage, part of the TLC227X, is shown in Figure 8.1.

The RRO characteristic is achieved in the construction of the op amp output stage. A totem pole design that has upper and lower output transistors is used, and the output transistors are a complementary pair. Each transistor in the pair is a "self-locking" type of transistor operating in the common-source mode. Consider the p-channel output transistor; as long as this transistor has a drain-source resistance it forms a voltage divider with the load resistance. When the load is a very large resistor or if the output current flow is very small, the voltage drop across the output transistor can be neglected. Output current flows through the

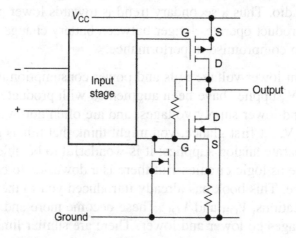

Figure 8.1
Rail-to-Rail Output (RRO) Stage

output transistor, and because current drops a voltage (V_{DS}) across the drain source resistor, the output voltage swing is reduced. The voltage drop subtracts from the power supply voltage, reducing the output voltage to less than RRO.

8.2.2 Dynamic Range

Dynamic range is affected by the V_{OH} and V_{OL} specifications of the op amp, but is a broader subject. Start with the maximum output voltage swing, $V_{OUT(MAX)}$. This output voltage swing is defined as the maximum output voltage the op amp can achieve (V_{OH}) minus the minimum output voltage the op amp can achieve (V_{OL}). V_{OH} and V_{OL} are easily obtainable from an op amp IC data sheet. This yields Equation 8.1:

$$V_{OUT(MAX)} = V_{OH(MIN)} - V_{OL(MAX)} \tag{8.1}$$

Equation 8.1 can be used to illustrate the role that power supply voltage plays in limiting the dynamic range. $V_{OH(MIN)}$ is the most positive power supply voltage minus the voltage drop across the upper output transistor; thus $V_{OH(MIN)}$ is directly proportional to the most positive power supply voltage. For any op amp, the output voltage swing is directly proportional to the power supply voltage; thus, in the same op amp, the dynamic range is directly proportional to the power supply voltage.

An op amp may have V_{OH} and V_{OL} specifications that are close to its power supply rails, but not necessarily equal. This is because the output transistors are real-world devices that have some voltage drop. So there will never be a true output rail-to-rail op amp, unless it is an op amp which has internal DC−DC converters to boost its internal rails. This is a lot of trouble to go to just to achieve a perfect rail-to-rail specification. Careful system design can achieve the required performance without resorting to such extreme measures. As an op amp has smaller V_{OH} and V_{OL} specifications, the semiconductor design challenges get greater and greater, which can lead to increased power consumption and susceptibility to latch-up.

8.2.3 Input Common-Mode Range

Just as there are limitations on the output voltage swing of an op amp, there are limitations on the input voltage range that may be applied to it. This can be troublesome, particularly if the input source is referenced to ground, and has small amplitude. Fortunately, there are such things as true rail-to-rail input (RRI) op amps. There are trade-offs associated with them, however, and it may be sufficient to use an op amp that includes either ground or positive power in its common-mode range, but not both.

Figure 8.2 shows the simplified input circuitry for an op amp that can go to the ground rail (but not positive power rail). The PNP input transistors are biased by the emitter current source. If the positive input is connected to ground bias current still flows and the transistor stays active.

An op amp with an NPN input stage works in a similar way around the positive supply rail. It can sense voltages close to V_{CC} and maybe slightly above V_{CC}, but it will not work when it is within 1.5 V of ground. The solution for this problem is to include parallel input circuits as shown in Figure 8.3.

There are both PNP and NPN differential amplifiers used in the input stages of the RRI op amp; thus the RRI op amp can operate above and below the power supply voltage. As Figure 8.3 shows, the parallel input stages can be made in bipolar or MOS technology. Inclusion of complementary differential input amplifiers achieves V_{ICR} exceeding the power supply limits, but there is a penalty to pay in input bias current, input offset voltage, and distortion. This has serious implications for low-level DC coupled systems.

The input stages operate in three different ranges:

- When the input voltage ranges from about -0.2 V to 1 V, the PNP differential amplifier is active and the NPN differential amplifier is cut off.
- When the input voltage ranges from about 1 V to ($V_{CC} - 1$ V), both the NPN and PNP differential amplifiers are active.
- When the input voltage ranges from about ($V_{CC} - 1$ V) to ($V_{CC} + 0.2$ V), the NPN differential amplifier is active and the PNP differential amplifier is cut off.

Figure 8.4 shows the input bias current and input offset voltage as a function of the input common-mode voltage.

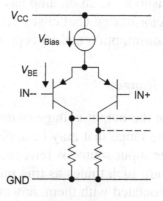

Figure 8.2

Input Circuit of a Non-Rail-to-Rail Input Op Amp

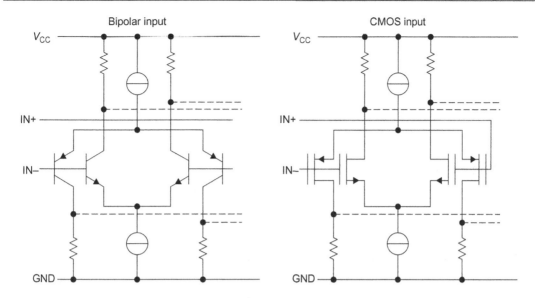

Figure 8.3
Input Circuit of a Rail-to-Rail Input Op Amp

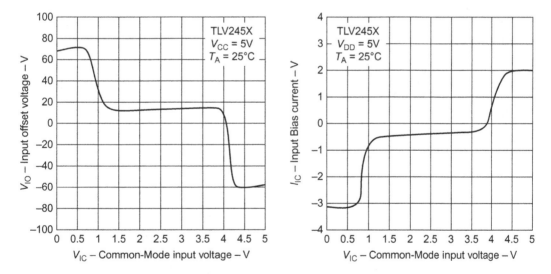

Figure 8.4
Input Offset Voltage and Bias Current Changes with Input Common-Mode Voltage

RRO op amps cannot drive heavy loads and maintain their RRO capability because of the voltage drop across the output transistors. Load resistance or output current is a test condition when the measurement of an op amp's output voltage swing is made. The size of the load resistor or output current is a measure of the op amp's ability to retain its RRO capability while sourcing or sinking an output current. When selecting an RRO op amp, you must consider the load resistance or output current required because these conditions control the output voltage swing.

When an op amp is made that has RRI and RRO capability, it is called a rail-to-rail input/output (RRIO) op amp.

8.2.4 Signal-to-Noise Ratio

Noise sets a limit on the information and signals that can be handled by a system. The ability of an amplifier, receiver, or other device to discern a signal is degraded by noise. Noise mixed with the incoming signal, noise generated by the op amp, resistor noise, and power supply noise ultimately determine the size of the signal that can be recovered and measured.

Noise fluctuates randomly over a period of time, so instantaneous signal or noise levels do not describe the situation adequately. Averages over a long period (root mean square or RMS) are used to describe both the signal and the noise. Signal-to-noise ratio (SNR) was initially established as a measure of the quality of the signal that exists in the presence of noise. This SNR was a power ratio, and it was established at the output of a circuit. The SNR that we are interested in is a voltage ratio because the impedance is constant, and it is established at the input to the op amp. This means that all noise voltages, including resistor noise voltage, must be calculated in RMS volts at the op amp input. The SNR is given in Equation 8.2:

$$SNR = 20 \log_{10}\left(\frac{V_{SIGNAL}}{V_{NOISE}}\right) \tag{8.2}$$

A good starting point is to test the analog signal chain with a terminated input, whether that termination is to ground, V_{REF}, or some other common-mode point. If there is a characteristic input impedance, such as 50 Ω, it should be included in the termination. If all goes well, the level of noise induced by the signal chain should be very low. Careful attention to layout and decoupling techniques is important for low-level signals.

Most signals are established by a transducer, which is a device that senses a change in a variable and converts that change into a voltage change. Transducers also convert some of their physical surroundings into a noise voltage that is combined with the signal. Noise from the physical surroundings of the transducer, unless its nature is well known, is almost impossible to separate from the transducer signal.

When transducers are connected to the electronics, cabling picks up noise and crosstalk from other signals, and some transducers such as thermocouples can pick up noise from the connecting junctions. Thus, the signal is never clean as it enters the electronics.

Transducers often have a very small output voltage swing, so when the transducer output voltage swing is converted to least significant bits (LSBs) the noise voltage should be very small compared to an LSB. Consider a temperature transducer that has a 10 mV swing over its range. When the transducer output voltage swing is considered to be the full-scale voltage (FSV) of an ADC, the LSB is very small, as shown in Equation 8.3 for a 12-bit (N) ADC:

$$\mathrm{LSB} = \frac{\mathrm{FSV}}{2^N} = \frac{10\,\mathrm{mV}}{2^{12}} = \frac{10\,\mathrm{mV}}{4096} = 2.44\,\mu\mathrm{V} \tag{8.3}$$

The op amp for this application must be a very-low-noise op amp because an op amp with a $20\,\mathrm{nV}/\sqrt{\mathrm{Hz}}$ equivalent input noise voltage and a bandwidth of 4 MHz contributes $40\,\mu\mathrm{V}$ of noise. This high noise contribution is why extensive filtering and "optimally" low bandwidth are desirable in the input stages of some electronic systems. If there is power supply noise, some of that noise passes through the op amp to its input. The power supply noise is divided by the power supply rejection ratio, but there is always a residual noise component of the power supply on the op amp input, as shown in Equation 8.4, where k_{SVR} is 60 dB:

$$V_{\mathrm{PS(INPUT)}} = \frac{V_{\mathrm{PS}}}{k_{\mathrm{SVR}}} = \frac{10\,\mathrm{mV}}{1000} = 10\,\mu\mathrm{V} \tag{8.4}$$

When interfaced with a data converter, the noise level ultimately determines how many bits of the digital code are meaningful; it does no good to purchase a data converter with a large number of bits when the level of noise in the analog signal chain makes several of the least significant bits irrelevant. If this is the case, you can:

- replace amplifiers in the signal chain with lower noise devices
- employ averaging over a large number of samples to reduce the contribution of noise
- use a data converter with fewer bits to save money, if the noise level is already the lowest that can be obtained, or is acceptable to system requirements.

8.3 Summary

It is extremely hard to achieve large dynamic range when the application is limited to a low power supply voltage. In an attempt to approach the dynamic range obtained by ± 15 V power supply designs, the new op amp designs put increased emphasis on the output voltage swing. However, the lower supply voltages have placed severe limitations on dynamic range, and attempts by semiconductor manufacturers to lower the V_{OH} and V_{OL} limits have regained only a fraction of that dynamic range, putting the burden on the user to be very careful in selecting the operating range of their circuits.

RRI op amps are able to work with transducers connected to the power supply rails. As long as the AC component of the transducer output voltage does not exceed the input common-mode range of the op amp, the design is reliable. RRI op amps are troubled by distortion introduced by the change in bias current, input offset voltage, and gain, but their contribution to the system's signal handling capability is invaluable. RRO op amps yield the highest output voltage swing of any series of op amps. But beware: RRO op amps are specified at a load resistance or current, and the output voltage swing decreases dramatically when the load resistance or current is increased. RRIO op amps contain the input and output features of RRI and RRO op amps. They also contain the drawbacks of both features.

The final thing to be considered is that low power supply voltage invariably means single-supply design, and single-supply design is tougher than split-supply design. Remember to get the two sets of data points, put them in simultaneous equations, solve for the slope and intercept, select the circuit configuration, and calculate the component values. Digital-to-analog converters (DACs) are a little different because you have to account for the polarity of the current, but their design generally follows the same procedure.

CHAPTER 9

Extreme Applications

9.1 Introduction

Not every circuit is destined to be used in a nice environment. Some are destined to go into extreme environments, or applications where failure is simply not an option. While military op amps have been available for a long time for deployment in temperatures that range from -55 to $125°C$, military environments are no longer the most severe. Space, oil exploration (downhole), and geothermal environments have much greater temperature extremes, as well as potentially damaging mechanical vibration and shock requirements. Space, medical, and automotive applications demand high levels of reliability. Service is impossible for space electronics. Medical devices can put a patient's life in danger if they fail. Automotive applications can cause traffic accidents if they do not work properly. This chapter is intended to be an introduction to the topic: the techniques here will be a starting point for extreme design, not a guarantee of operation of a given system in an extreme environment. That is left to the designer, who must adhere strictly to a set of qualification tests developed by the company for which they work. A system is suitable for extreme environments only after such qualification tests have been passed.

9.2 Temperature

This section will focus mainly on high-temperature design, because that is the environment the author is most familiar with. When dealing with high temperatures, some of the op amp parameters you care about will degrade, and the trend will accelerate on some parameters as the temperature increases. Put another way, the degradation of parameters is not necessarily linear, and the temperature at which a given parameter begins to accelerate towards unacceptable may differ between parameters. Curiously, other parameters may remain unchanged. Table 9.1 is based on a real-world op amp which is available in commercial and high-temperature grades; however, the exact op amp will not be mentioned because its specifications are subject to change by the manufacturer.

Table 9.1: Comparison of Ambient and High-Temperature Parameters

Parameter	Commercial Grade	High Temperature	Units
V_{OS}	± 125	± 260	µV
$\Delta V_{OS}/°C$	1.5	2	µV/°C
I_B	± 200	± 250	nA
I_{OS}	± 150	± 150	nA
e_n	1.1	1.1	nV/ QUOTE
I_n	1.7	1.7	pA/ QUOTE
CMRR	120	113	dB
AOL	110	110	dB
GBW	45	45	MHz
SR	27	27	V/µS
t_s	400	580	nS
THD + N	0.000015	0.000015	%
V_{OH}	(V +) − 0.2	(V +) − 0.2	V
V_{OL}	(V −) + 0.2	(V −) + 0.2	V
I_{SC}	+ 30/ − 45	+ 30/ − 45	mA
Z_O	5	5	Ω
IQ	6	7.5	mA
PSRR	140	140	dB

For a full understanding of these op amp parameters, look at Appendix A. Here, I will briefly describe what these parameters mean to you for this particular op amp. The news is definitely not all bad!

9.2.1 Noise

The specifications related to noise − e_n, input noise current (I_n), and total harmonic distortion plus noise (THD + N) − do not change at high temperature. A low-noise op amp, in this case, remains a low-noise op amp (which probably drove the selection of this part in the first place). Parameters related to conducted sources of noise such as the power supply rejection ratio (PSRR) and common-mode rejection ratio (CMRR) are not affected or barely affected − another lucky break.

9.2.2 Speed

The parameters related to speed − the gain bandwidth product (GBW) and slew rate (SR) − do not change at high temperature. While this is not a particularly high-speed op amp, it retains the speed it had at room temperature. In some cases, I have actually seen the bandwidth of a high-speed op amp increase with high temperature, albeit by just a few percent. But it is one of those rare cases where the performance of the part may actually be better at high temperature than it is at room temperature.

9.2.3 Output Drive and Stage

There is some more good news for this device: its output stage appears unaffected by high temperature. The high- and low-level output voltages (V_{OH} and V_{OL}), short-circuit output current (I_{SC}), and output impedance (Z_O) all are the same at high temperature. This is good news if you plan to drive a heavy load, or if you need a rail-to-rail output. But be careful: most rail-to-rail op amps degrade under load conditions. It is always wise to check the graphs in the data sheet, and doubly so for parameters like these that may interact.

9.2.4 So, What Degrades at High Temperature?

The supply current increases at high temperatures, which will cause battery life to decrease in battery-operated systems. An op amp which was not targeted for precision DC transducer applications has higher values of offset voltage and input bias current at high temperature. Beware of the input bias current: it can vary by orders of magnitude at temperature for some devices. You ignore this effect at the risk of a non-functional product.

The final parameter that degrades at temperature is the most important: LIFE. An integrated circuit (IC) operating continuously at high temperature is slowly destroying itself. ICs operating at high temperatures can have lifetimes in the thousands or even hundreds of hours, orders of magnitude less than the same device operated at ambient temperature. A mission profile that realistically projects the amount of time for which the circuitry will be operated at elevated temperatures is imperative. A downhole application is a good example. The amount of time the electronics is actually subjected to elevated temperature may be relatively short in relation to the amount of time the tool is used downhole. Temperature inside the hole rises (and falls) slowly, not reaching maximum temperature until the electronics is actually at maximum depth. On the other hand, a space application may see very rapid temperature fluctuations that create thermal shock conditions, which can rapidly degrade device packaging.

9.2.5 Final Parameter Comments

Just because these particular parameters varied the way they did for this particular op amp, does not mean these variations will hold true for every op amp. It is your responsibility to do a "what if" analysis for every parameter you have any doubts. Designing for worst case is always a good idea, and this is doubly so for any extreme application, high temperature in particular. You may have to use a commercial grade component for prototyping purposes, because the high-temperature versions are much more expensive. The good news is that the high-temperature version of the part is probably fabricated from a commercial grade die that has been screened for high-temperature performance, so there is a very good chance it will work properly at high temperatures for short periods. In such cases the difference between a commercial grade device and a high-temperature device is packaging: the commercial grade die is screened for high-temperature performance, and then placed in a high-temperature package.

Experts at companies specializing in these applications can be a valuable source of information. Often, a list of parts that have survived harsh environments in the past is available, and use of those parts is preferred, if at all possible.

9.3 Packaging

Another aspect of extreme applications that needs to be addressed has to do with packaging. This includes the IC itself, the board on which it is mounted, the solder which attaches it to the board, and other aspects. All of these must be seriously analyzed and understood, or the circuitry may not survive in the extreme environment.

9.3.1 The Integrated Circuit Itself

Many IC manufacturers are starting to supply products for extreme-temperature applications. The portfolios, however, are still tiny, and costs may be very high. The ICs may be packaged in specialized packages that do not match standard packages, or may have different pinouts. Fortunately for you, the vast majority of ICs can operate at extreme temperatures, at least for short periods.

The silicon die is seldom the limiting factor in high-temperature operation. I have encountered only one IC that does not work at high temperature; all of the others that I have tested do, except those with intentional over-temperature shutdown "protection". Voltage regulators are the most notorious for this feature. There is usually no need for it, and it severely limits the number of voltage regulators that can be used.

Problems begin with the connection of the bond wire to the die. High temperatures accelerate undesirable metal migration, eventually leading to failure at the connection. High currents exaggerate the problem even more. Proper high-temperature bonding techniques are a must at high temperatures. Related to this is the selection of bond wire. Specialized alloys may be necessary for high-temperature operation. Both these alloys and the bonding techniques are closely guarded trade secrets, and you will have to work carefully with IC manufacturers to make sure proper attention is paid to both issues.

9.3.2 The Integrated Circuit Package

One of the fundamental problems in high-temperature design is the thermal expansion characteristics of the various materials used. The silicon die itself has a thermal expansion characteristic, the package enclosing it has a thermal expansion characteristic, the bond wires have thermal expansion characteristics (affecting their length), and so on. Adhesives have thermal breakdown characteristics. Plastics may out-gas corrosive or alkali components. Repeated temperature cycles allow water to penetrate the packages. Cold temperatures make things brittle, and high vibration and shock may break something mechanically. All of this is a nightmare not only for IC manufacturers, but also for designers.

The best, and least expensive, solution may be to characterize a commercial part in a plastic package for its lifespan in the extended temperature environment. If a given "mission" is limited to a few hundred or a few thousand hours and the environment is relatively well understood, a commercial or industrial grade IC in a plastic package may well survive in the vast majority of cases. But an IC destined for space where the mission is years long must be packaged in high-temperature packaging if there is to be any hope of survival.

High-temperature packaging is usually ceramic, with gold leads. Ceramic packaging has come into use for extreme applications from a military background, where ceramic through-hole and flatpack packaging has been used for decades. Both alternatives may be undesirable for applications with limited board space. New surface-mount ceramic packages designed for extreme temperatures are becoming available, but not all ICs are available in these packages. You may be forced to buy dies from the IC manufacturer and have them packaged by a third party. In extreme cases, you may have to resort to die-harvesting services, where the die is physically removed from its plastic package and repackaged into high-temperature packages. It is extremely important to work with a partner who has experience in high-temperature applications and knows the requirements for the package and bond wires.

9.3.3 Connecting the Integrated Circuit

Now that you have your high-temperature IC, how do you place it onto a board? If you are used to soldering it onto an FR-4 printed circuit board (PCB) with Sn60Pb40 solder, you have just wasted your money. The board will disintegrate to ashes at high temperature, and your IC will fall off because the solder melts! High-temperature operation requires a complete rethinking of PCB materials and solders.

9.3.3.1 Printed Circuit Board Design

The material of choice for high-temperature PCBs is traditionally glass-reinforced polyimide resin, but this material is porous and will absorb water. If a board gets wet, it is probably scrap. Gold plating is usually done on the traces, but it requires a nickel barrier between the copper and gold, or copper will dissolve into the gold, causing brittle solder joints.

Thermal expansion and contraction also take place with PCBs, making the selection of hole size, via size and trace widths larger than most PCB designers are accustomed to working with. You may have considerable "pushback" from PCB layout departments, but it is important to hold your ground. Surface-mount components help with the situation by minimizing the number of through-holes which act as barriers on all layers of the board. The mass of surface-mount components must be minimized, or they may tend to come off the board under high shock and vibration conditions. They also may have to be oriented in a common direction to avoid mechanical stress along an axis that may flex. As boards expand and contract, they may tend to break connections at pads and vias. Teardrops or "necking down" should be used to avoid stress points where traces connect to pads and vias.

9.3.3.2 Solders

Solders for high-temperature applications have been migrating towards or high melting point (HMP) solders. The best of these appears to be Sn05Pb92.5Ag2.5, which has a melting point of 280°C, 536°F. These solders require new tips; it is unwise to contaminate new solder joints with a mixture of HMP and older solders. HMP solder has also been described as acting like "silly putty that melts only once", a very good description coming from someone who has been frustrated by it. I have been forced to use a lead-free solder, Sn96Ag04, on more than one occasion to get a solder joint that behaves more like an Sn60Pb40 solder. It certainly is harder to flow through small holes, and requires a new set of skills to work with. Just because you can solder does not mean you can solder well, and it certainly does not mean you can work with HMP solders. It is usually best to leave HMP soldering to technicians, assemblers, and third party houses that can handle it.

9.3.3.3 Adhesives and Other Methods of Component Retention

A board for an extreme environment is never assembled completely when the components are soldered. Many techniques are needed and used for securing components to a board that will be subjected to extreme levels of vibration and stress. These techniques must also be suitable for high-temperature use. Adhesives are usually high-temperature epoxies, but care must be taken that their thermal expansion and contraction characteristics do not induce stresses on the board or IC.

Toroidal inductors and transformers are commonly used, because they are usually custom wound on high-temperature cores. Retaining them on a PCB usually involves drilling extra non-plated through-holes on the board, and lashing down with high-temperature lashing, retaining with a screw and top plate, or packaging the toroid in a housing which already has leads. Any of these techniques must ensure that vibration/movement in all possible axes is taken into account. They may be combined with the use of adhesive as well.

Heatsinking is difficult in some applications, such as downhole where space is at a premium. Still, there is a very good heatsink located very close to the electronics: the tool itself. Just remember that it is more of a heat spreader than a heatsink because it is at elevated temperature in the hole. But there is usually a supply of drilling fluid which will also act to draw heat away. Finding a way to thermally attach a component to the metal of the tool is a challenge, and every company probably has an approved method. Space applications cannot count on airflow past a heatsink because of the vacuum of space. Heat will flow out of the heatsink, but at a much reduced rate. Heatsinking is certainly not an option in implantable medical devices, so low-power design is imperative. Chances are, the implantable device is already designed to be at low power owing to battery life constraints.

Adhesives and potting materials are often used in high-shock environments to retain components that do come loose, so they will not be distributed to critical areas as mechanical nuisances or shorts. The board from which they came may or may not be disabled by the loose component, if it was a decoupling capacitor for example; but the loose component must not disable anything else! High-temperature adhesives and potting must also be selected for low conductivity if high voltages are present.

9.4 When Failure Is Not an Option

The Mythbuster motto is "failure is always an option". That may be true, but often our job as designers is to run counter to that truism as much as possible. There are no set design techniques to ensure that failure is never an option, because some sets of circumstances will cause any circuit to malfunction. The only way to counter this is to make failure happen only after a specific set of circumstances applies: the default mode if something fails is something safe.

Extreme applications include a class of applications where failure would jeopardize life or property. Medical equipment falls into this category: if failure of an implantable device would jeopardize a patient's life, or if malfunction of hospital equipment could hurt or kill a patient. Property damage could include things like malfunctioning circuitry causing an oil spill in a remote location. Legal implications alone from such events could result in a company going out of business and its owners being imprisoned if found guilty of negligence. If you are a licensed professional engineer (PE), you could also be held personally liable.

There are many techniques that can be used to reduce the probability of failure. A good place to start is to examine the power supply/battery. A reliable circuit without reliable power is one that is doomed to fail. Reliable power is the cornerstone of reliable electronics, and must be examined first.

After that, where do you go? Worst case design should already be standard practice: never design a circuit around typical values. Beyond that:

- If an electronic design must work under any set of circumstances, develop a mindset of "what if". What if power fails: will the system default to a state of least damage? A variable pulse-rate pacemaker should revert to single frequency, not shutting off, if at all possible; a patient would be impaired, but not dead. I remember working on a large wire wrap computer board. In order to prove its self-diagnostic capabilities, a manager would cut a random wire, and see how long it would take the team to find the failure. Such exercises might seem annoying, but at least one can rule out or compensate for some potential failure modes.
- Beware of transistor-related specifications. A lot of designers like to be clever and use transistors, displaying their prowess at biasing. Some transistor parameters, however, vary by orders of magnitude over temperature extremes. When I utilize such circuits, if my transistor switch or amplifier stage will not work with biasing resistors from 100Ω to $1\,M\Omega$, I redesign until it does. And do not assume you are immune if you only use op amps. Input bias current is a specification that can vary unexpectedly over temperature. Be wary of it.

- Prototype everything. Simulations are starting points, not ending points.
- Innovation and parts pushing the state of the art are not good ideas. Sticking to older, proven technology is smarter.
- Redundancy is a powerful technique. Two power supplies sharing current, but each one capable of providing full load current independently, will greatly increase reliability. Two completely redundant signal chains will allow a measurement to be made even if one signal chain goes down.
- Rely on a set of circumstances that is counter to randomness. I recall a valve control system for an underwater wellhead that would only actuate valves when a timer operated in a narrow range of frequencies, charging a capacitor at just the right rate. Any other set of circumstances, and the value remained tightly closed, preventing environmental disaster — pretty clever, and simple! Other techniques might make critical actions happen only when a set of prime numbers is sent to a system, sequences of prime numbers being uncommon in nature.
- Exhaustive testing is imperative. When a specific test regimen is in place, following it to the letter will minimize liability to managers and outside agencies. When a test regimen is not already in place, establish one appropriate to the level of risk involved, and have it approved by corporate legal departments before selling a single product.
- Do not accept arbitrary, overaggressive schedules. When substantial risk of injury and/or property damage is at stake, it is foolish to allow yourself to be rushed. If something bad happens because you did not take the time to investigate, you are the one who will be liable, especially if you have a PE in front of your name. The very managers who were pushing you to meet the schedule will be running for the hills in the event of problems. If you are not prepared financially to join them, just say "No"! Remember the Challenger disaster: engineers were afraid to speak up; lives were lost.

Some very sobering thoughts, I know. But when you are designing such systems, the buck stops with you. Take ownership of the responsibility you took on.

9.5 When It Has to Work for a Really Long Time

Another extreme application has nothing to do with temperature, shock, or vibration extremes. Extreme duration in time is another very challenging area of design. It may be a space probe like a Mars Rover, an implantable medical device, a monitor on a remote pipeline, or a detonator on a mine floating in the ocean. In each case, the mission can be measured in years or even decades. In each case, the system has to work over the entire mission time. And in some cases, like a mine detonator, most of the work is done at the end of the device's lifespan.

This produces extreme requirements on the power system, whether it is battery, inductively coupled super capacitors, nuclear, solar, or some other technique. Chances are that aspect of the design has been done for you, and you do not have to worry about it. But you can greatly enhance your value as a designer by minimizing the load you draw from such critical, limited resources.

Extremely low power consumption is a cooperative effort between power supply/ battery engineers, digital designers, software/firmware coders, and those designing the analog signal chain. Many compromises and trade-offs will be involved, but the primary way to save power is by incorporating power switching into the design. The analog signal chain, for example, is only turned on when needed to acquire a signal from a sensor. The clock frequency of the system is adjusted based on need.

In the aforementioned mine detonator, the clock ran at a very low frequency, and then at an intermediate frequency when it detected motion, then at full frequency to characterize "friend or foe". Clock circuits are also power hogs, and a special low-current consumption clock oscillator was constructed. Op amps were very low power, and design compromises were made to accommodate the extra noise, larger offset, and lower bandwidth from such op amps. The result was a system that consumed power only when it needed to, and was extremely low power at all other times. When average power was considered, system lifetime could be measured in decades or even centuries, yet instantaneous power when running complex algorithms was comparable to that for any commercial product running data acquisition.

One area of concern to you as a designer of extremely low power consumption systems is the switching/monitoring circuitry that has to be "on" for a long time. It is very important that everything be taken into consideration, even the leakage currents through electrostatic discharge (ESD) protection diodes. Power monitoring/ switching circuitry can become quite involved. When in doubt, do not discount small microcontrollers if multiple inputs can turn on a system. Extremely low power consumption microcontrollers exist that can painlessly handle different sets of circumstances from more than one input. Everything must be rigorously tested and prototyped.

9.6 Conclusions

This chapter has no figures, no equations, and only one table. It has only a lot of text, a lot of that being the voice of long and hard experience with systems that have failed in unexpected, catastrophic ways. It is sad that we tend to remember our failures much better than our successes. After decades in this industry, I have had my share of failures on the road to successful designs, have seen others fail, and have picked up the pieces after other designers failed. So I hope that some of my insights have helped. But there is no methodology, no formula, no magic bullet that will create an absolutely reliable design. If there were, it would have been developed by now. High-reliability design for extreme applications is an exercise in discerning and managing risk areas, i.e. those areas in a design that are most likely to fail, addressing those, and then looking for the next most likely failure modes and addressing those, until the design is reliable enough or has failure modes low enough to allow deployment in the given situation. It is not an area for the inexperienced, nor is it one for those who like to push the envelope in new features and use the newest, fanciest parts. It can, nevertheless, be very rewarding to point to a system that has worked as designed for years or decades and say that you had a hand in creating something that will last. So many pieces of high-tech equipment are obsolete almost as they leave the factory, discarded in a year or two for a newer model. Reliable systems can be a point of pride or a selling point on a résumé.

Voltage Regulation

10.1 Introduction

In Chapter 3, Table 3.1 introduced a comprehensive listing of cases. In between the cases for inverting and non-inverting gain was a line on the table when op amp gain $m = 0$ and offset b is positive, negative, or zero. These fall into the category of voltage regulators, and have their own set of interesting challenges.

I have included some design utilities on the companion website, which are discussed at the end of this chapter.

10.2 Regulator Cases

10.2.1 Virtual Ground: $b = 0$

Let's start with the simplest case, $b = 0$. This is the ground reference point of a circuit, and requires no active circuitry. However, active virtual grounds are sold, and find a niche market in line driving applications. Rather than simply leave you with a simplistic "$b = 0 =$ ground connection" answer, I will refer you to Appendix C, which discusses proper circuit board grounding techniques.

10.2.2 Positive and Negative Voltage Regulators: $b > 0$, $b < 0$

The remaining cases are voltage regulators. When there is a positive offset ($b > 0$), the circuit is a positive voltage regulator, and when there is a negative offset ($b < 0$), the circuit is a negative voltage regulator. These circuits are designed for power applications, to provide power to other circuitry. AC gain in regulator circuits ($m \neq 0$) is undesirable because it is ripple voltage. You need to be aware of ripple voltages in circuits, and the power supply rejection ratio of the op amp will tell you how much the ripple voltage will be rejected. Be especially careful with high-gain op amp circuits powering switching regulators, as the ripple voltage may be amplified if the power supply rejection ratio of the op amp is not large enough.

Op Amps for Everyone.
DOI: http://dx.doi.org/10.1016/B978-0-12-391495-8.00010-6

Regulators can be linear or switching (utilizing an internal oscillator to perform voltage-level conversion). Entire volumes have been written pertaining to both types of regulator; this chapter makes no effort to go into the details of regulator design other than to introduce the concept of the op amp as a feedback device controlling the output voltage level.

An interesting subset of voltage regulators are the voltage references. These can be as simple as a zener diode, or they can have the same architecture as a linear voltage regulator, just optimized for voltage output accuracy. There is no reason why a voltage reference cannot be used to power low-power circuitry, as its architecture is identical to that of a linear regulator. However, that is not the intended application of the part. Using it at a substantial portion of its maximum rated load may compromise its precision specification.

10.3 Make or Buy?

Voltage regulators are ubiquitous devices in many applications. They are available in a variety of output voltage, power levels, and packages. Many of them have multiple outputs. With all the work already done, what is the motivation for you to go to the trouble to design your own?

With highly integrated voltage regulators also comes some baggage:

- Downhole and geothermal applications operate at high temperatures. The vast majority of integrated regulator integrated circuits (ICs) have overtemperature protection, designed to keep the IC from burning up if too much load current is drawn when the device is already at an elevated temperature. Designers for these applications have extensive experience in derating components, and are used to components being operated in a manner that limits their usable life. Therefore, overtemperature shutdown is an unwanted feature that renders a voltage regulator IC unusable.
- Cellular telephone base stations are installed at tens of thousands of remote locations, in small buildings or enclosures. Because a base station is a complex and power-hungry device, the buildings and enclosures require air conditioning, especially in hot climates. A common failure of air-conditioning systems is loss of coolant, which will cause the equipment to overheat and shut down. While this is acceptable — there is no way to harden the complex digital circuitry — many regulator ICs and their close cousins the power supply "bricks" or modules have an undesirable characteristic, in that they require power cycling or a manual "reset" to recover from fault conditions. This is unacceptable for systems in remote locations, which require regulators to automatically recover

from fault conditions. So a designer of these systems might elect to design their own regulator.

- A switching regulator may not be available in the frequency range needed. In the case of the legendary GE/RCA Superadio 3 radios, a DC–DC converter was used to step up 9 V DC to a highly regulated 15 V for the tuning voltage. Not only was the highly regulated voltage an unusual requirement, but the application forced the switching frequency of the regulator above the amplitude modulation (AM) band, to eliminate interference. Hobbyists have found it extremely difficult to reproduce this requirement with commercially available ICs now that the original regulator IC has been discontinued. Discrete designs using a regulator IC in combination with a precision 15 V reference design have fulfilled the requirement, although the solutions are bulky compared to the original IC.

There are many other reasons why you might opt to build your own rather than buy an IC to do the job: cost, lead time, the availability of spare op amps/reference voltages in the system, and so on.

10.4 Linear Regulators

Figure 10.1 shows how a linear voltage regulator functions. Both the linear and switching regulators operate similarly as far as the control circuitry (the op amp feedback loop) is concerned. The linear regulator, however, is more simple and will be presented in some detail. Please refer to Figure 10.1 for the following discussion.

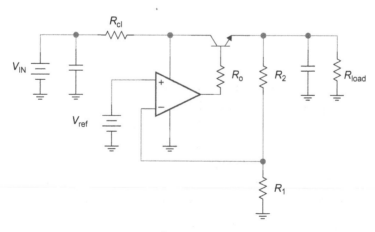

Figure 10.1
Voltage Regulator Operation

The input voltage, V_{IN}, is assumed to be a voltage above the voltage supplied to R_{load}. Furthermore, it is assumed to be large enough that the transistor above the op amp can be biased into its linear region. This difference between the input voltage and output voltage can never be zero, and is referred to as the dropout voltage. Dropout voltages can range from a few tenths of a volt to several volts. If you decide to do your own design, low dropout voltage is probably not going to be easy.

The input voltage is also assumed to be a bit messy, having some ripple or noise on it. Therefore, it is filtered with decoupling capacitors, which are selected to have maximum effect on the ripple. Spend the time necessary to characterize this ripple and choose capacitors that are best at eliminating it; this is time well spent because it makes the job of the regulator that much easier. Some regulator designs fail simply because they do not have sufficient bypassing, and break into oscillation.

A current-limiting resistor, R_{cl}, has been added to this circuit and is good protection against a short-circuit on the output in this simple configuration. It will be the element that fuses and protects the rest of the circuit should a short occur. R_{cl} is often placed elsewhere in the circuit when a regulator IC is used, and will be part of a less destructive protection scheme. Suffice it to say that with simplicity come some trade-offs!

The op amp is the control loop element. It is powered by the input voltage, as is a voltage reference. This reference can be supplied external to the circuit, generated off a precision zener diode, or even be a precision voltage reference IC. It is supplied to the non-inverting input of the op amp, which should have precision DC specifications, especially if used over a wide temperature range.

If the reader takes only one thing from this section, chapter, or even book, it should be this:

> *An op amp will do whatever it can to the output voltage in order to make the voltage on its two inputs the same voltage.*

That statement above can be used to derive everything that has ever been written about op amps, and every application that uses them. In the case of a voltage regulator, this statement explains the op amp as a control element in the regulator. R_2 and R_1 form a voltage divider monitoring the output voltage (emitter of the transistor in this example). The inverting input of the op amp is supplied with the voltage-divided output voltage, and will adjust the bias on the base of the transistor to balance the two inputs; in other words, make the voltage divider match the reference voltage, V_{REF}.

Now for a brief explanation of the transistor. It is called a "pass" transistor in the world of power supply regulation. It can be just about any transistor, bipolar or even a field-effect transistor (FET). It should be able to dissipate enough power to supply the output power required by the circuit, plus dissipate its own power caused by the voltage it is expected to drop, plus the load current. Therefore, the pass transistor is the element that takes the beating in a regulator circuit. Many lower power regulators have the pass transistor built in, while others provide a way for using an external pass transistor that can dissipate much more power than an internal transistor can. In the case of this example, the op amp output has been buffered with a resistor R_o to limit the base current in the pass transistor. This is another protection scheme: if the pass transistor were to short, this would prevent damage to the op amp output.

The response time of the op amp in the control loop is limited only by the bandwidth of the op amp. In some cases, this may result in problems. The op amp response to transients can be speeded up by addition of a capacitor across R_2, or slowed down by a capacitor across R_1. This requires you to do a bit of experimentation. At any rate, the output voltage should be decoupled by capacitors chosen to reduce ripple coming through the regulator. Of course, R_{load} is a representation of the circuit load. The circuit load is not necessarily resistive. The output decoupling capacitor already makes it capacitive; if the load has substantial inductive characteristics, it may be wise to put a diode across the pass transistor to protect the circuit in case of back-spikes from the load. In normal circuit operation, the diode should be reverse biased, only conducting if V_{OUT} exceeds V_{IN}.

10.5 Switching Power Supplies

The switching regulator shown in Figure 10.2 is a variation of the circuit of Figure 10.1. At first glance, this schematic may seem formidable. Upon closer examination, however, it becomes clear that the voltage divider network from Figure 10.1 consisting of R_2 and R_1 is still there. The output connects to an FB pin 3 of the IC. If you consult the data sheet for the device, it goes to the inverting input of an "error comparator" in Figure 10.1, nothing more than an op amp. The non-inverting input of the op amp goes to a 1.5 V reference generated on the IC. All of the sudden, design of this part of the switching power supply circuit is beginning to look a whole lot like that of the linear voltage regulator, and with good reason. That is because it is the same principle exactly, only instead of controlling gate bias to the pass transistor directly, it is now indirectly controlling the amount of time the transistor is "on" and, in this case, magnetizing the inductor.

This particular IC has yet another easily understood op amp circuit, one which is common to a great many voltage regulator ICs: the current sense circuit. If the reader looks at the circuit in Figure 10.3, the CS input is on pin 8, and on the low

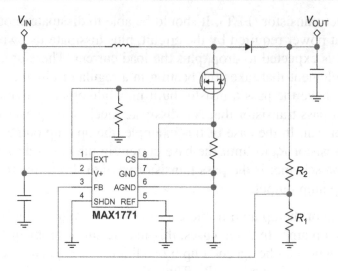

Figure 10.2
Switching Regulator

side of the transistor switch. So instead of the transistor switching the inductor directly to ground, it is taken through this resistor, which is usually a pretty small value in the tens or hundreds of milliohms. It therefore generates a small voltage, which is used by yet another op amp (used as a comparator) inside the IC. A 0.1 V reference inside this particular IC determines the trip point: when the CS input exceeds 0.1 V, the IC shuts down, detecting an overcurrent condition.

10.6 A Companion Circuit

Consider the case when a voltage regulator fails, allowing unregulated input voltage to be applied to the load. This is almost always catastrophic; after all, if raw unregulated voltage would not cause a problem, why not use it in the first place? Fortunately, it is easy to add an overvoltage protection circuit to greatly reduce the likelihood of this happening. Granted, there are no absolutes, and there are always circumstances where the regulator and the protection circuitry fail at the same time. But playing a game of odds, anything short of a lightning stroke or direct connection to line voltage will probably not produce such a failure. If both the regulator and overvoltage protection circuitry fail at the same time, the whole system should come under suspicion because something really dramatic has happened.

There are two types of overvoltage protection circuits: one that opens the voltage path and does not allow voltage to pass to the load, and a shorting or "crowbar"

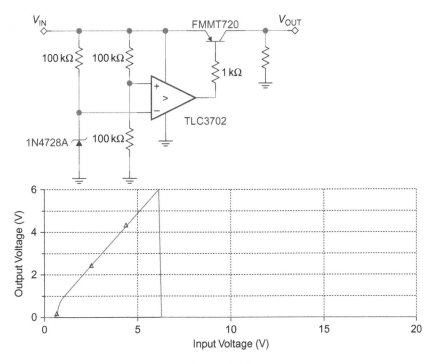

Figure 10.3
Overvoltage Protection Circuit

type of protection that shorts overvoltage to ground, thus blowing any fusible element on the input voltage. This, of course, is a non-recoverable condition which may be appropriate for some applications, but why design a non-recoverable circuit when it is just as easy to design a recoverable one?

Consider the circuit of Figure 10.4. It is designed to protect the circuitry powered by a 5 V regulator, which is powered by a 20 V input. The load current is 250 mA. The input voltage V_{IN} comes from the voltage regulator output, and the load is applied to V_{OUT} (represented by a resistor). This figure looks very much like Figure 10.1, but there are some crucial differences. First, it is the input voltage that is sampled instead of the output voltage. Second, the op amp has been replaced by a comparator.

The pass transistor is a PNP switching type, optimized for low collector–emitter voltage drop and high current. Small package size is achieved even though the transistor is passing high current; its voltage drop is so low that the overall power dissipated in the transistor is low. Looking at the connections to the comparator:

- A zener diode is used to set a voltage reference at the inverting input.

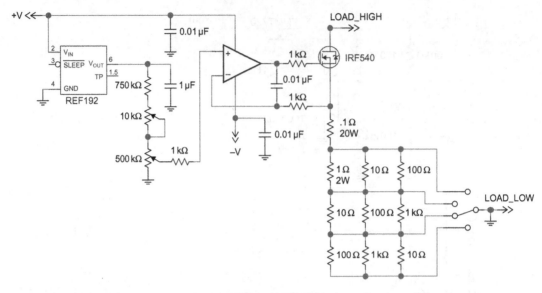

Figure 10.4
Active Load

- The input voltage is monitored by a voltage divider on the non-inverting input.
- The output is connected through a current-limiting resistor to the base of the pass transistor.

Looking at the DC sweep, there is a region of operation where the comparator is not under power. Fortunately, that is not required for the pass transistor to be "on". All that is required is that the base be low. As the input voltage continues to increase, the comparator turns on with its output in the low state (the non-inverting input is lower than the inverting input). A 1N4728 is a 3.3 V zener diode, which is adequate for this circuit to function. Fine adjustments can certainly be made by adjusting resistor values and choosing a different zener. As the input voltage reaches 5 V, the voltage applied to the non-inverting input of the comparator is still below the threshold set by the zener diode on the inverting input, so the comparator output is still low and the pass transistor conducts. As the input voltage continues to rise, however, the comparator inputs reach equality, and the comparator switches its output high, turning off the pass transistor. V_{OUT} goes to zero, and current goes to zero. There is a small amount of leakage, but it is inconsequential. The comparator output continues to rise along with the input voltage. This may be a concern if there is a large input voltage, but the base resistance can be increased at the expense of forward voltage drop if need be.

A similar technique could be used for overtemperature protection, with the comparator set up to monitor temperature on the inverting input instead of the input voltage. A second pass transistor may not even be needed, if you are clever in using open collector type comparators!

10.7 Another Companion Circuit

This section is offered as a way to test the reader's comprehension of the previous section, or at least as a way to give the reader a very useful test circuit.

Take the circuit of Figure 10.1, and make the following changes:

- Break the connection of the transistor to the input voltage.
- Allow adjustment of the reference voltage.
- Define the load in terms of a resistance – 0.1Ω, 1Ω, 10Ω, or 100Ω – plus the "on resistance" of the pass transistor.

What is left is the active load circuit of Figure 10.4, which is a very useful way of testing power supply circuits. When power supplies are tested with a power resistor, the load current rises and falls along with the power supply output voltage. When tested with this circuit – an adaptation of a circuit from Maxim Semiconductor – the load current is independent of the power supply voltage.

Consider the case where:

- The voltage applied between LOAD_HIGH and LOAD_LOW is 1 V.
- The switch is in the position indicated, for a total resistance of 10Ω (plus the on resistance of the IRF540).
- The 10 kΩ pot is adjusted to zero.
- The 500 kΩ pot is adjusted to 500 kΩ.

Therefore:

- The inverting and non-inverting inputs of the op amp are both at 1 V.
- The op amp output has reached its positive rail; the transistor is turned on as "hard" as it can be, causing it to effectively be a short, and drop no voltage.
- Therefore, the 10Ω resistor drops all of the load, or 100 mA.

Now, increase the voltage to 10 V:

- The op amp inputs are still both at 1 V.
- The op amp output is 4.732 V; this is probably not a useful number for the reader, but definitely accomplishing something at the transistor gate.
- The transistor is now dropping the applied voltage by 9 V, leaving 1 V at the 10Ω resistor.

- The total load current is (10 V − 9 V)/10Ω = 100 mA. The current is the same as before.

Therefore, the load current is independent of the power supply voltage. One further modification: adjust the 500 kΩ potentiometer to 250 kΩ:

- Both op amp inputs are now at 0.5 V.
- The op amp output is now at 4.2 V.
- The transistor is now dropping 9.5 V.
- The total load current is now (10 V − 9.5 V)/10Ω = 50 mA.

In practice, the 500 kΩ pot is the fine current adjustment, and the rotary switch is the current range. Be very careful of part wattage; be sure to heatsink the transistor and power resistors properly. This circuit is an inexpensive and handy addition to the test bench of any power supply designer. A side benefit is that this circuit can be set to any load current and frees you from having to have a stock of power resistors to use as dummy loads.

10.8 Design Aid

The voltage regulator circuits shown in Figures 10.1 and 10.2 require a resistor divider circuit to set the output voltage to the desired level. I have written a JavaScript calculator (Figure 10.5) to aid in the design of this network. To use it, you have to know the value of the internal voltage reference and the output voltage. There is a drop-down box for resistor sequence and resistor scale. The calculator may produce more than one option for resistors.

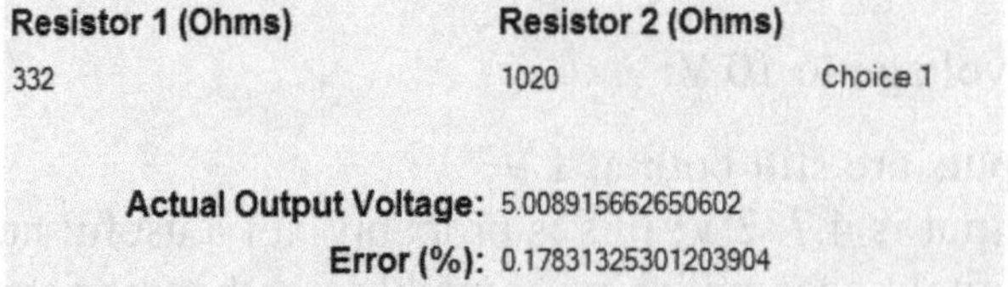

Figure 10.5
Voltage Regulator Calculator

Please remember:

- The voltage regulator topology may be linear or switching; the switching topology may differ quite a bit from the figures presented here and on the calculator, but the calculator is concerned only with the feedback network needed to set the output voltage, so the exact topology is not important to its operation. Only the correct value of internal reference is important.
- The regulator may be sensitive to resistor scale. If the data sheet is making recommendations for R_1, stay within a decade of it.
- This calculator is not intended to address the complexities of bypassing on the input, output, or reference. Neither does it address inductor values or transistor selection for switching regulators. This can be tricky, and if you do not know what you are doing, leave those aspects of the design to experts.
- Some regulators may have feedback networks to select undervoltage shutdown. This calculator can be used, in some cases, to design those networks as well. As long as an internal reference voltage is known, the calculator will work.

10.9 Conclusions

Even an op amp designer who does not consider themselves fluent in power supply design can utilize many of their skills in designing parts of the feedback loop. Rather than being intimidated by power supply design, you should embrace it, because you already have most of the basic skills. You can make custom voltage regulator circuits that are specific to the system's needs, and add as much or as little sophistication as necessary to the application − all with just basic op amp design skills!

CHAPTER 11

Other Applications

11.1 Introduction

This chapter discusses some other common and interesting op amp applications. Three applications will be discussed: interfacing digital-to-analog converters (DACs) to loads, oscillators, and power boosters. None of these is a complex enough topic to warrant its own chapter, but combined here they should round out your designer's toolkit and make a complete chapter.

11.2 Interfacing Digital-to-Analog Converters to Loads

The techniques of Chapter 3 are applicable to transducer to op amp circuitry to analog-to-digital converters (ADCs), and now this chapter will introduce the reverse process: interfacing a DAC to a load. The design techniques are similar, but with the difference that these devices use current outputs, not voltage.

Design Example

An op amp is used to interface a DAC with an actuator (Figure 11.1). Although you could easily convert the current output to a voltage output by merely supplying a load resistor, it is best to directly utilize the output current. DAC accuracy is optimized for current output. Often, the DAC provides an integrated op amp on the chip, with its parameters optimized for current input mode. This is shown in Figure 11.2.

Consider this example: the DAC output sinks current from the power supply I_{OUT} $_{(ZEROS)} = -1$ mA to $I_{OUT(ONES)} = -2$ mA at an output compliance of 4.33 V. The output requirement is a voltage swing of $V_{IN1} = 1$ V to $V_{IN2} = 4$ V, with a load resistance of 100 kΩ. A 5 V power supply is available. Assume that the design uses 5% resistors. The DAC is connected to the input of the amplifier (Figure 11.2), so its output current swing is I_{IN}. The input/output range is $I_{IN1} = -1$ mA at $V_{OUT1} = 1$ V and $I_{IN2} = -2$ mA at $V_{OUT2} = 4$ V. These data points are substituted into Equation 11.1. Do not worry about the sign of m or b, because it is determined

Op Amps for Everyone.
DOI: http://dx.doi.org/10.1016/B978-0-12-391495-8.00011-8

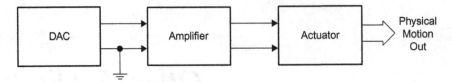

Figure 11.1
Digital Control System

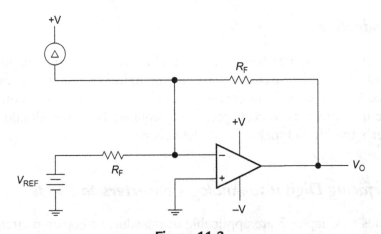

Figure 11.2
DAC Current Sink to Actuator Interface Circuit

by the math, and it is substituted into the equation that determines the transfer
equation. The transfer function for the current sink DAC is given in Equation 11.1:

$$V_{OUT} = I_{IN}m + b \qquad (11.1)$$

The simultaneous equations are given below:

$$1 = -m + b \qquad (11.2)$$

$$4 = -2 + b \qquad (11.3)$$

From these equations we find that b = −2 and m = −3. The slope and intercept
values are substituted into Equation 11.1 to get Equation 11.4:

$$V_{OUT} = -I_{IN}(-m) - b = mI_{IN} - b \qquad (11.4)$$

The current equation for the circuit shown in Figure 11.2 is given below as
Equation 11.5, and after algebraic manipulation it becomes Equation 11.6:

$$\frac{V_{OUT}}{R_f} = I_{IN} - \frac{V_{REF}}{R_g} \qquad (11.5)$$

$$V_{OUT} = I_{IN}R_f - V_{REF}\frac{R_f}{R_g} \qquad (11.6)$$

Comparing terms between Equations 11.4 and 11.6 enables the extraction of m and b:

$$R_f + |m| = 3 \qquad (11.7)$$

$$|b| = V_{REF}\frac{R_f}{R_g} \qquad (11.8)$$

$$\frac{R_f}{R_g} = \frac{2}{5} \qquad (11.9)$$

These equations are written in terms of mA and kΩ. In the final schematic of this example (Figure 11.2), R_g is selected as 51 kΩ, so $R_f = 20$ kΩ. When $I_{IN} = 2$ mA, the compliance of the DAC is 0.0 V.

This example presented one possible DAC output interface; it is hoped that you can adapt the technique to your situation. The design procedure is similar to those for the cases of Chapter 3, but utilizing the current output of the DAC makes it sufficiently different to warrant this section.

11.3 Op Amp Oscillators

When talking about op amp oscillators, you need to unlearn everything you have previously learned about op amp stability criteria; in other words, you now want to purposely design a circuit which, by its very nature, is unstable.

At this point, I could launch back into the theory of feedback design, but to get to the point: the best way of making an op amp unstable is to use positive – not negative – feedback. Negative feedback as used up to this point has operated to limit output voltage excursions in the op amp circuit. Positive feedback, on the other hand, reinforces the output voltage excursions, exaggerating and amplifying them, in phase, until the op amp output is saturated. This leads to an important decision: whether to use op amps or to use a very similar, but different component – the comparator (Figure 11.3).

Both circuits in Figure 11.3 were designed to give an output of approximately 10 kHz. One fact emerged very quickly: the technique used to determine filter cut-off points does not work. If it had worked, one would have expected both circuits to operate with a resistance/capacitance (RC) circuit comprised of 4.42 k and 3600 pF

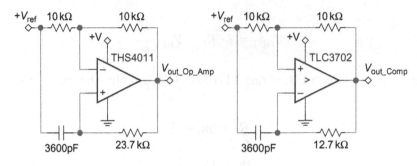

Figure 11.3
Oscillator Schematics

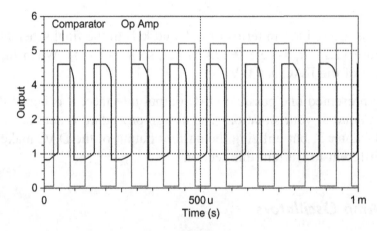

Figure 11.4
Oscillator Outputs

capacitors. The resistor value had to be "tweaked" considerably to slow both of these circuits down, and the resistors are different for the op amp and the comparator. But this is a small matter compared to the suitability of the device.

So, what is the difference between using an op amp and a comparator in this application? A look at Figure 11.4 gives a very quick answer. It is apparent that the op amp output is degraded compared to the comparator. After reading this volume, you should at this point be familiar with the V_{OL} and V_{OH} specifications of an op amp. In this case, those limitations clamp the voltage excursion of the op amp output to about 0.8–4.5 V. But that is not the end of the story. The leading edge of the waveform is fairly decent, but the trailing edges show a different picture: the op amp output stage is tending towards latch-up, and takes some time to recover. If the voltage swing of the op amp was not a convincing enough reason to use a

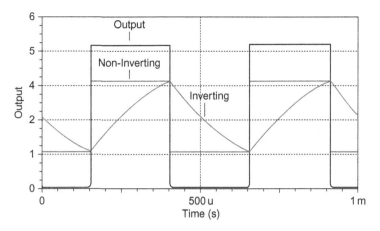

Figure 11.5
Comparator Oscillator Analysis

comparator, the latch-up tendency surely is! It points to a potential failure of the circuit due to latch-up stresses being repeatedly applied. The comparator, on the other hand, has an output that is designed to switch from rail to rail. This makes it ideal for use in an oscillator circuit.

Next, let's delve into what makes this circuit oscillate. Taking op amps out of consideration, the comparator circuit is analyzed in Figure 11.5.

The non-inverting input is easy to understand: it is a voltage divider from the output to the reference voltage. Initially, with the comparator output low, the voltage divider presents about 1 V to the non-inverting input. When the comparator output switches high, the voltage divider presents about 4 V to the non-inverting input.

When the inverting input is analyzed, it is an RC charge and discharge circuit. Initially, the capacitor is charged to the reference voltage, but since the output is low, the capacitor goes into a discharge cycle until it reaches a voltage equal to that on the non-inverting input. At that point, the output goes high, and the capacitor is charged until it reaches the higher voltage on the non-inverting input, at which time the comparator output switches back low. And that is how the circuit operates: the capacitor is continually charging and discharging to reach the voltage on the non-inverting input.

Many designers want a sinusoidal oscillation, not a square-wave oscillation. That can be easily accomplished by filtering the circuit above, but not necessarily the output! You should instead filter the voltage that appears on the inverting input, because a triangle wave's harmonics are lower than a square wave's harmonics, and the voltage swing is within the voltage swing limits of a unity gain op amp stage.

This may sound a bit counterintuitive, to use the inverting input of the comparator as its "output", but you will be fine as long as the op amp stage does not load the inverting input to any degree. If there is any doubt, use the output of the comparator instead.

11.4 Hybrid Amplifiers and Power Boosters

If you are careful, it is possible to create a composite amplifier whose important characteristics are better than those of either individual op amp.

The circuit in Figure 11.6 shows a composite amplifier constructed by an OPA277 and an OPA512. The OPA512 is used as a power output buffer in the feedback loop of the OPA277. Characteristics are tabulated in Table 11.1.

The OPA512 has the highest slew rate and therefore is operated within a local closed loop. The slower OPA277 is operated within the outer loop. The 47 pF capacitor provides a small amount of phase shift to help stabilize the system. The resulting performance of the compound amplifier shows that the front-end characteristics of the OPA277 are joined with the ± 35 V at 10 A drive current capabilities out of the OPA512. The slew rate of the OPA277 is 0.8 V/μs. That slew

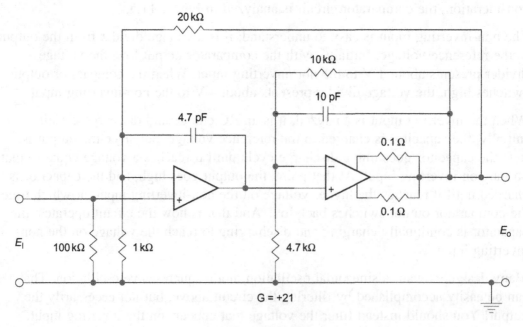

Figure 11.6
Composite Op Amp

Table 11.1: Composite Amplifier

Parameter	OPA277	OPA512	Compound
V_{OS}	20µV	6 mV	20µV
Drift	0.15µV/°C	65µV/°C	0.15µV/°C
I_B	1 nA	30 nA	1 nA
CMRR	130 dB	100 dB	130 dB
V_{OUT}	± 13 V	± 35 V	± 35 V
I_{OUT}	5 mA	10 A	10 A
SR	0.8 V/µS	2.5 V/µS	2.4 V/µS

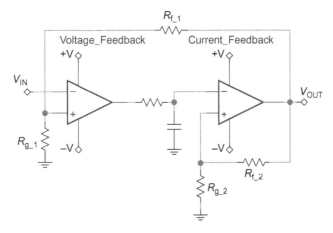

Figure 11.7
Composite High-Frequency Op Amp

rate is gained up times three in the OPA512 so that there is an effective slew rate for the compound amplifier of 2.4 V/µs.

High-speed hybrid amplifiers can be constructed as shown in Figure 11.7 so they take advantage of the speed and power output of current feedback op amps.

In this configuration, you should use the values recommended in the data sheet for R_{f_2} and R_{g_2}. This particular schematic shows the current feedback amplifier used in a gain mode; it is definitely possible to use it in unity gain mode by leaving out R_{g_2}. Remembering, however, that the least stable op amp configuration is unity gain, I usually operate the current feedback amplifier at a gain of 2 just to slow it down a bit. The RC network in between the two stages is there to slow things down in case the circuit breaks into oscillation. You can also slow things down by placing a capacitor in parallel with R_{f_1}. Remember that the values of R_{f_1} and R_{g_1} will determine the stage gain, but the values of R_{f_2} and R_{g_2} will determine the maximum voltage swing possible out of the voltage feedback amplifier. V_{OL} and V_{OH} of the current feedback amplifier will determine the voltage swing of the stage. You may have to play around with this circuit a bit to get it to operate; there are a lot of things to balance and compromise to obtain acceptable performance.

What if more power is needed? It is possible to parallel op amps to increase the output power level. Figure 11.8 shows the technique. The output of the voltage feedback amplifier has been connected to an RC low-pass filter as before. But this time, two current feedback amplifiers are used. Their outputs are connected to small series resistors, between 1Ω and 5Ω – the exact value again determined by a bit of experimentation. The series resistors are there to prevent one current feedback amplifier from driving the output of the other; they compensate for small mismatches in the components. You should check the wattage; they may have to be power resistors. Like anything else in analog design, there are trade-offs. In this case, the output voltage swing of the stage is reduced because of the series

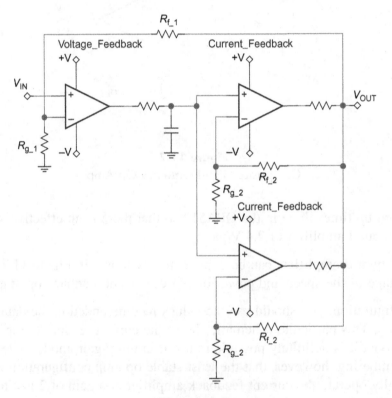

Figure 11.8
Composite High-Speed Op Amp with Power Boosting

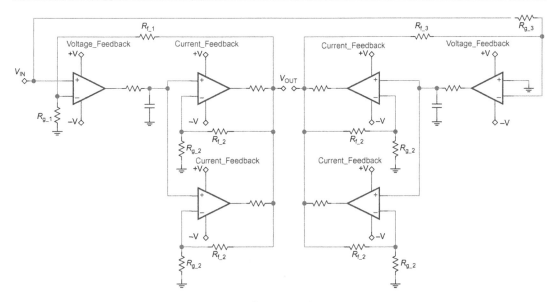

Figure 11.9
High-Speed Bridged Hybrid Amplifier

matching resistors. Because power is V^2/R, this is a fairly serious compromise, but a necessary one. It can be combated by raising the power supply voltage, or by operating the circuit in a bridged configuration (see Figure 11.9).

The output V_{OUT} is a differential output, with two voltages equal and opposite in polarity, thereby doubling the output voltage swing and quadrupling the output power. There are many techniques for inverting one polarity of the signal while not inverting the other. The technique used here is to balance the non-inverting gain formed by R_{f_1} and R_{g_1} with the inverting gain formed by R_{f_3} and R_{g_3}. Since there is a magnitude difference of 1 between them, the resistors cannot be the same value. But very good approximations can be made to give:

$$\left| \frac{-R_{f\_3}}{R_{g\_3}} \right| = 1 + \frac{R_{f\_1}}{R_{g\_1}} \tag{11.10}$$

For example, for $R_{f\_3}/R_{g\_3} = 4$ and $1 + R_{f\_1}/R_{g\_1} = 4$, the resistors could be:

- $R_{f\_3} = 10.2$ k
- $R_{g\_3} = 2.55$ k
- $R_{f\_1} = 20$ k
- $R_{g\_1} = 10$ k

Placing these values into Equation 11.10 yields:

$$\left| \frac{-10,200}{2500} \right| = 1 + \frac{20,000}{10,000} \qquad (11.11)$$

Both sides of Equation 11.11 are exactly 4, so the polarities of the two stages would be opposite, and the magnitudes equal, and the circuit perfectly balanced. This assumes that the current feedback amplifiers are operated at unity gain (R_{g_2} not used).

11.5 Conclusions

I hope that these three applications were fun to look at and easy to understand. They are not ones you will encounter every day, but this brief introduction should help you to understand them when they come along.

CHAPTER 12

Manufacturer Design Aids

12.1 Introduction

This chapter discusses some of the online resources available from semiconductor manufacturers. I include some of my experience with them and review them a bit. These online resources are subject to change at any time by the manufacturers, of course, and they own the names. I may seem a bit critical at times of some of the limitations and shortcomings, but I am genuinely impressed by the quality of free online offerings and the generosity of the companies who put them up there.

I do have some general comments regarding these online tools. Often, they contain hooks to encourage you to buy the "full, unlimited version" from a third party software author. The free versions of the programs may be feature limited, time limited, node limited, or have other limitations that "cripple" the software. You have to learn to live within the limitations. It is possible to get very creative in working around some of the limitations to the point where you rarely, if ever, have to worry about them. The goal, of course, is to do some serious engineering work and all of these design aids are focused in that direction. I cannot provide a user's manual for these tools. Some of them are quite complex and it would take dozens or hundreds of pages to delve into them.

Many manufacturers have minor web-based calculators and design aids. There are hundreds of them, so I cannot include all of them here. Most are very simple and intuitive. There are also engineering forums on some manufacturer sites, some sponsored by magazines, etc. I am a member of most of them and you may see my contributions on them. I am a regular columnist on EE Web.

A word about Spice models: most semiconductor manufacturers provide Spice models. They vary widely in quality and compatibility. Invariably the new device, or the device you most want to use, will not have a model, or the model will not import correctly. I have had to get creative in identifying parts from competitors that have similar parameters. But the quality of Spice models is a continuing headache for designers, one that semiconductor manufacturers should really address.

DOI: http://dx.doi.org/10.1016/B978-0-12-391495-8.00012-X

12.2 Texas Instruments Tina-TI

This is one of my favorites, and certainly among the best free tools out there. A screen shot is shown in Figure 12.1.

Shown here is a simulation of a double notch Twin-T filter designed to reject both 50 and 60 Hz. Tina-TI will only work if one op amp is included in the schematic. Fortunately for you, an "ideal" op amp is included which you can use as a starting point. If you want to analyze a circuit that does not contain an op amp, you can simply drop the ideal op amp into the design at some point as a non-inverting buffer where it will not load the circuit.

The Spice Macro tab is loaded with TI Spice macros. You cannot create a new macromodel circuit with the Spice file from another manufacturer. The File open tab has an example circuits selection containing the majority of TI products. Many of these example circuits were actually designed by me during my tenure at TI, or by John Bishop, another op amp expert whom I worked with. They can be used as a starting point for more complex designs.

The basic tab is where you will get 90% of what you need to build a schematic: the power sources, ground, resistors, and capacitors. Semiconductors are generic, like the ideal op amp. Like some other editors, placing the mouse cursor near an unconnected pin changes the context to place connections, and the user can lock the context of the cursor on the top menu bar.

Pros

- Tina-TI simulations have neatly organized options and are intuitive.
- A very nice feature of Tina-TI is the ability to right-click the simulation window results, and paste those results directly into the schematic file. You will see this feature utilized in the example circuits.
- It does not seem to be node or time limited, unlike the Analog Devices version of NI Multisim (see Section 12.5).

Cons

- I have found Tina to be a bit slow, particularly in transient analyses.
- Tina is a commercial product, and you will be asked whether you want to buy the full version when you attempt to access a limited feature. I am a licensed user of the full version, and I will be frank and tell you that the registration process for the licensed version is cumbersome, slow, and prone to rejecting legitimate customers.
- It is not a US product, so there may be a time zone difference when trying to get support. Email is your best bet.

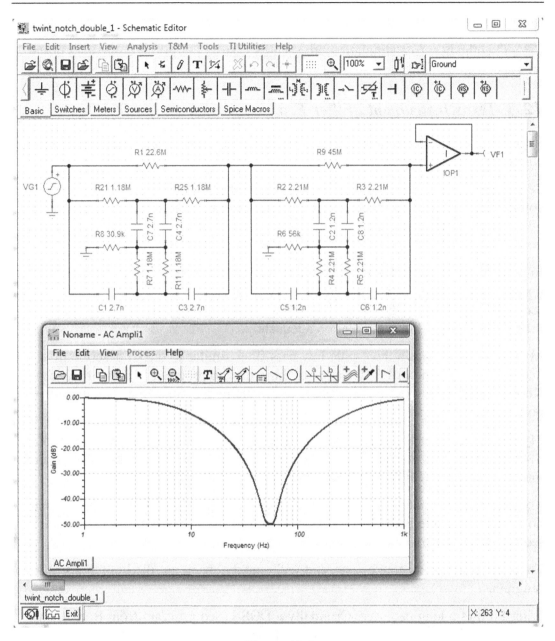

Figure 12.1
Tina-TI Simulation

Bottom Line

Tina-TI is the best free simulation program by a large margin. It deserves a place on your desktop.

12.3 Texas Instruments Filter Pro

First, a few general comments about free filter design programs (and by extension commercial filter design programs).

Let me preface this discussion by saying that neither Filter Pro nor any other free or commercial product is perfect. All of the filter design programs, Filter Pro included, will produce schematics of usable filter circuits (for the most part). They may not, however, produce an optimized solution using the fewest op amps.

In doing these reviews, I was interested in testing the absolute limits of the programs: how well do they handle the very toughest cases — the worst case scenarios? If I do not talk about all of the test cases for a particular program, it handled the ones I do not mention adequately.

My first test case was a three-pole low-pass filter similar to ones I describe in Section 6.5.1 (Chapter 6) (offering something that no program I know of can do). Knowing that, how close do they come to producing a simple filter based on two pole stages?

My second test case is a high Q bandpass filter: how well does the program produce a solution that does not fall apart in Monte Carlo analysis [the weakness of general multiple-feedback (MFB) bandpass filter implementations]?

My third case is a notch filter: how well does it handle Monte Carlo analysis, and how manufacturable is the filter when you look at the bill of materials — does it require matched parts and how many?

These are some pretty tough tests, and will quickly check out the limitations of a program.

Filter Pro has a long pedigree. Originally a DOS-based program from Burr Brown, TI acquired the rights to it when they purchased Burr Brown. John Bishop and I spearheaded the effort to bring the program into the Windows platform. He did the coding and I did the algorithms. Two revisions were produced, the first being a low-pass-only version, and the second adding high-pass, differential filters and a few other new features. An unreleased beta version added bandpass, notch, and many other features; however, TI went in a different direction after John and I left the company.

The new version of Filter Pro takes a lot of ideas from competing commercial products such as Filter Wiz Pro, particularly the user interface. I have, of course, been a big fan of Filter Pro since rescuing it from DOS oblivion, so I decided to put this new version through its paces.

The opening screen of Filter Pro is friendly enough. I selected low pass (Figure 12.2).

On the next screen I entered the parameters for a three-pole low-pass filter: a cut-off frequency (− 3 dB) of 1 kHz, a 60 dB roll-off at 10 kHz, and unity gain (Figure 12.3).

Allowable passband ripple and the choice of entering gain as either dB or V/V is nice. The next screen gives a choice of filter response characteristic (Figure 12.4).

Filter Pro indicates four poles instead of three and two stages. An extra pole probably does not hurt anything. The previous versions of Filter Pro will default to

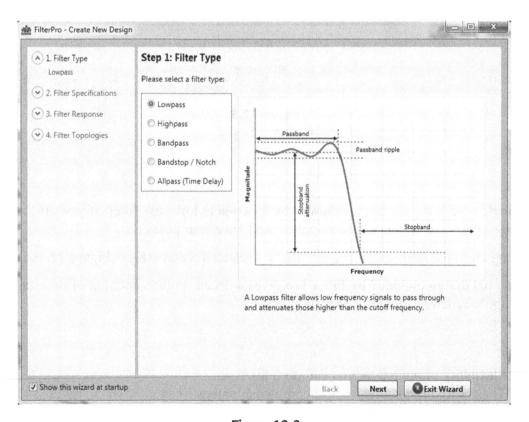

Figure 12.2
Filter Pro Opening Screen

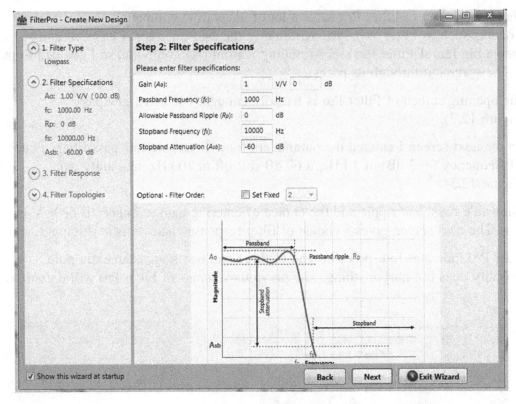

Figure 12.3
Filter Pro Step 2

a single-pole low-pass filter, followed by a two-pole low-pass filter. If you are going to use two op amps, you might as well have four poles.

Going on to the next screen (Figure 12.5), I selected Sallen–Key (Figure 12.6).

The final design provided by Filter Pro gives a decent implementation of two-stage, two-pole Sallen–Key low-pass response.

Pros

- Flexibility of options – you can do it all.
- The ability to select real-world component values.
- Fully differential filter option – this is the only program which offers this.

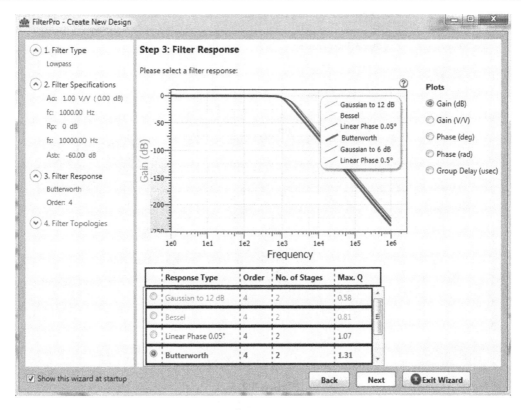

Figure 12.4
Filter Pro Step 3

Cons

- The notch filter produced by Filter Pro did not simulate in Tina-TI – TI's simulation program!
- Filter Pro has an option for E-96 capacitor values – something I have never seen for sale anywhere. You simply cannot buy E-96 capacitors.
- Exact component values are the default.

Bottom Line

If you can locate one of the previous versions of Filter Pro, definitely download it and use it. If not, this is still a very useful, intuitive program that deserves a place on your desktop. Simulate all designs that it produces, however, before committing to production.

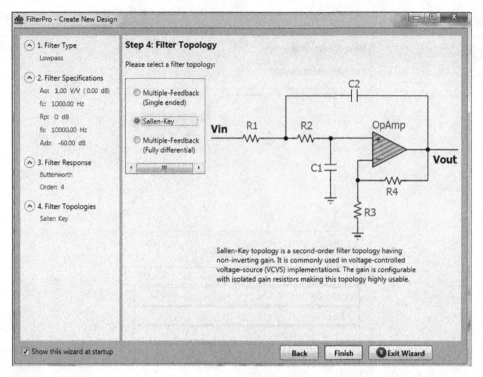

Figure 12.5
Filter Pro Step 4

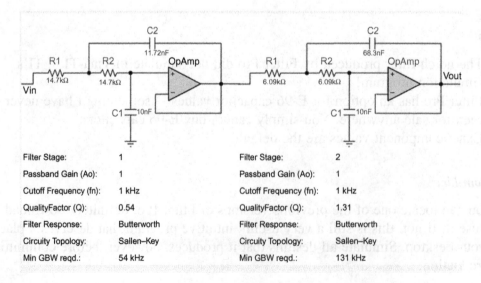

Figure 12.6
Filter Pro Final Design

12.4 *National Semiconductor/Texas Instruments Webench*

At the time of writing this book, the merger of Texas Instruments and National Semiconductor has been finalized, but there are still separate websites. National Semiconductor has maintained its own design aid – Webench – and I would expect it to be continued in its present form, although the user interface may be altered visually to match other TI tools.

Webench is intended to be something that Ron Mancini dreamed of: an all-encompassing design tool incorporating power supply, op amp gain, and filtering (and as many other design aids as possible) into one unified program. Many of these functions are included under the Webench umbrella; a tab can take you to op amp gain, another to filters, another to power supplies, etc. As this book is most interested in op amps, I will concentrate on them.

Sadly, the op amp gain section does not include offsets. It is very nice for basic inverting and non-inverting stages.

The filter design tab is available in the opening screen (Figure 12.7). Clicking it takes you to a response selection (Figure 12.8).

Select both the Filter Type and the Filter Performance Specification; the latter is mandatory. Not shown in these screen captures is a big red "Continue" arrow. Clicking it or Start Design has the same effect.

The next screen is where you start designing the low-pass filter. I entered the same parameters as I did in Filter Pro (Figure 12.9).

Figure 12.7
Webench Opening Screen

Figure 12.8
Webench Filter Design Section

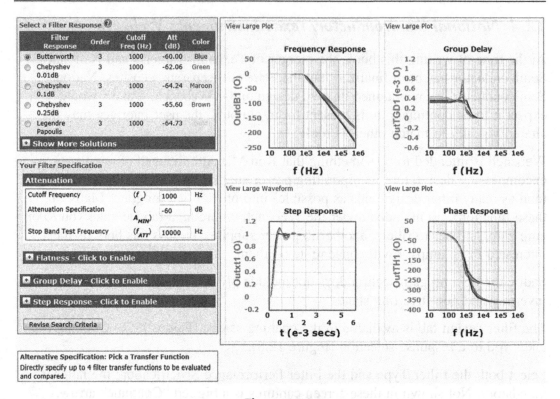

Figure 12.9
Webench Filter Parameter Entry

The next screen shows the final design (Figure 12.10).

Pros

• A unified development environment encompasses a lot of areas of design.
• This is the only program to offer op amp gain calculations.
• There are fewer steps to get to the final filter design than with Filter Pro.
• The filter order has been correctly identified as three.

Cons

• It curiously defaults to 0.5% resistors, but it is easy to switch using a drop-down menu.
• There does not appear to be an option for capacitor sequence, leaving you stuck with the E-24 sequence, some of which can be hard to obtain.
• This is a web-based tool only, so your designs are "in the cloud". Print out everything so you do not lose it!

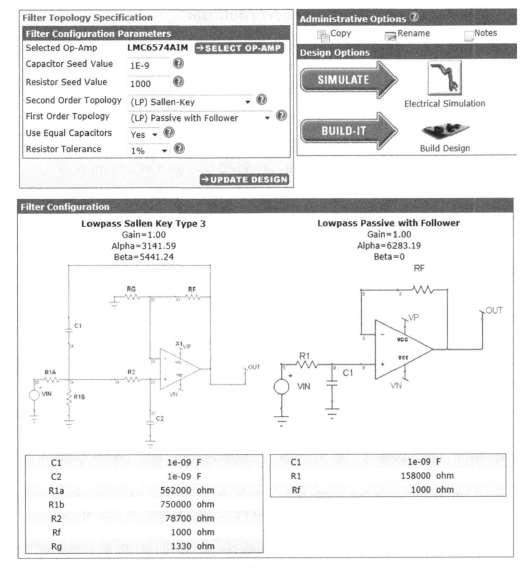

Figure 12.10
Webench Final Design

Bottom Line

Give it a try, there is nothing to dislike here. I have had issues with other sections of it; it is a massive effort involving many product lines, but the op amp/filter section is decent.

12.5 *Analog Devices Version of NI Multisim*

This is a very popular simulation program. Originally called Electronics Workbench, it gained a reputation for being intuitive, fast, and accurate. I will not include screen shots here, because you are either familiar with Multisim or not. It is very similar in look and feel to Tina-TI, and there are only minor variations regarding where things are located on the tabs and commands.

I have been a long-time user of Electronics Workbench, from its inception. It was my simulator of choice after PSpice was absorbed into the OrCad family of products and its price was inflated.

Pros

- It has wide industry acceptance, plus familiarity.
- I will give Multisim very high marks for including virtual Agilent and other test instruments in the simulation environment; this is absolutely wonderful, and they work like their real-world counterparts so well that lab set-up is a breeze, as is working on screen captures of the simulation schematic. I have little use for the LabView plug-ins currently available because we do not use it at my present employer, but I have gone through LabView Basic 1 and 2, and see how this is extremely useful for LabView users.
- It has better simulation of transient response than with Tina-TI.

Cons

- Multisim, unfortunately, has become a bit bloated of late, due in part to an attempt to incorporate board layout into the package. I have often found that Spice simulation programs are a bit cumbersome to use in an environment where you are aiming to produce a commercially viable board. Spice programs are packed with power sources, probes, and simulation hooks that do not translate into printed circuit board (PCB) symbols; PCB layout is packed with objects such as screws, stiffeners, and connectors that do not belong and are not included in simulation programs. Just my philosophy; perhaps Multisim has addressed these issues. But for a simulation-only program, choosing a section of a four-section op amp chip is an unwanted extra step.
- One feature I really miss on Multisim that is present on Tina is the ability to capture those simulation results into the schematic window. To do this in Multisim, I have to capture the area of the screen, import into a graphics program, trim, and save as a huge .BMP format. I can then "Place/graphics/picture" into the window. This is fairly labor intensive: right-clicking the results window is so much nicer.

- It is component limited to 50 components, and limited to the import of two custom components.
- It may be time limited to nine months. I cannot absolutely confirm that, because my current employer provided the full commercial version before that time limit. Given the price of the commercial product, purchasing it may not be an easy task through company channels. I suggest Tina-TI as an alternative if this limitation proves to be real.

Bottom Line

Download this if you are going to be working extensively with Analog Devices op amps. Also download if you are working at a company where LabView is used extensively, and/or you are certain that your employer is going to standardize on Multisim as the company standard for schematics and/or board layout.

12.6 Analog Devices OpAmp Error Budget

I found this utility while looking down Analog Devices' extensive list of utilities. It piqued my interest because it had a few of the advanced capabilities of the now discontinued OpAmp Pro (Figure 12.11). There is no support for offsets, unfortunately, but it should handle gain calculations quite well. Strangely, a few intermediate values overlay the right half of the screen.

It is the next screen that is really useful. Error sources are systematically entered, and a realistic approximation of circuit performance degradation appears. This was one of the most powerful features of OpAmp Pro, and I am glad to see it in some form on the web (Figure 12.12).

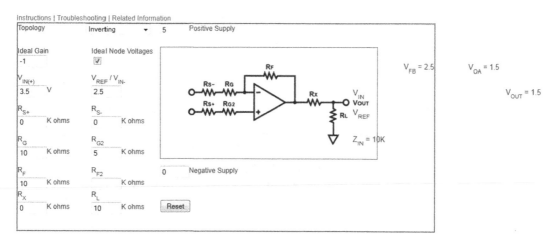

Figure 12.11
OpAmp Error Budget Top Box

Application Parameters						
Operating Temp., T_A	85	°C		Update		
Supply Variability (ripple+load reg.)	1	%				
Error Source	**Specification**	**Approx. Calculation**		**Absolute Error**	**Drift/Gain Error**	**Resolution Error**
Resistor Tolerance	0.1	%		2000 ppm		
Resistor Drift, TC_R	25 ppm /	~ (1/2 : noninv) $TC_R \times T_{DIFF}$			125 ppm	
Temp. difference, T_{DIFF}	°C 5	°C				
Nom. Open Loop Gain, A_{OL}	500	V/mV		6 ppm		
Min. Open Loop Gain	300	V/mV				4 ppm
Input Offset Voltage, V_{OSI}	0.2	mV	$V_{OSI} / (V_{IN}\text{-}V_{REF})$	400 ppm		
Input Offset Voltage Drift, V_{OSI_TC}	1.3 µV / °C	(2 : inv.) $V_{OSI\_TC} \times (T_A\text{-}25)$ / $(V_{IN}\text{-}V_{REF})$			156 ppm	
Bias Current, I_B - Source Imbalance Error	11 nA	$(I_B / (V_{IN}\text{-}V_{REF})) \times$ $(R_F\|(R_G+R_{S\text{-}}) - (R_{G2}+R_{S+}))$		0 ppm		
Bias Current Drift, I_{B_TC} - Source Imbalance Drift	N/S pA / °C	$(I_{B\_TC} \times (T_A\text{-}25) / (V_{IN}\text{-}V_{REF})) \times$ $(R_F\|(R_G+R_{S\text{-}}) - (R_{G2}+R_{S+}))$			0 ppm	
Offset Current, I_{OS} - Source Imbalance Error + Source Resistance Error	2 nA	$(I_{OS} / (V_{IN}\text{-}V_{REF})) \times$ $(3^*(R_F\|(R_G+R_{S\text{-}})) - (R_{G2}+R_{S+}))/2$		20 ppm		
Offset Current Drift, I_{OS_TC} - Source Imbalance Drift + Source Resistance Drift	N/S pA / °C	$(I_{OS\_TC} \times (T_A\text{-}25) / (V_{IN}\text{-}V_{REF})) \times$ $(3^*(R_F\|(R_G+R_{S\text{-}})) - (R_{G2}+R_{S+}))/2$			0 ppm	
Common Mode Rejection Ratio, CMRR	104 dB	(inv: (1+1/gain)×) $10^{-CMRR/20} \times$ $\| (V_++V_\text{-})/2 - (V_{S+}+V_{S\text{-}})/2 \| /$ $\| V_{IN}\text{-}V_{REF} \|$		31.5 ppm		
Power Supply Rejection Ratio, PSRR	120 dB	(inv: (1+1/gain)×) $10^{-PSRR/20} \times$ $(\| V_{S+}\text{-}V_{S+nom} \| +$ $\| V_{S\text{-}}\text{-}V_{S\text{-}nom} \|) / \| V_{IN}\text{-}V_{REF} \|$ $10^{-PSRR/20} \times$ SUP-VAR $\times$ $(V_{S+}\text{-}V_{S\text{-}}) / \| V_{IN}\text{-}V_{REF} \|$		0 ppm		0.05 ppm
Noise BW	0.1 - 100 Hz					11.5 ppm
Voltage noise, V_{NW}	15 nV/root-Hz	5 Corner freq Hz				
Current noise, I_{NW}	0.13 pA/root-Hz	5 Corner freq Hz				
Total resolution error						15.6 ppm
Total drift / gain error					281 ppm	
Total absolute + drift + resolution error				2760 ppm		

V 1.0.0 CPP3

Figure 12.12
OpAmp Error Budget Parameter Entry and Results

Pro

- The originality of the program, the fact that it gives functionality, you cannot find elsewhere.

Con

- Curious artifacts appear on the home screen.

12.7 Linear Technology LT Spice

LT Spice is another all-encompassing style of program that incorporates switching power supplies as well as op amps. I loaded an example circuit called "opamp" into the design window (Figure 12.13).

I had to manually add an "output" terminal to the circuit. When I simulated, it gave a very fast Bode plot, but a Bode plot that was completely blank. Right-clicking the blank window brought up another window where I could select the *V*(output) that I added. A maximize on just the simulation results, and the results are shown in Figure 12.14.

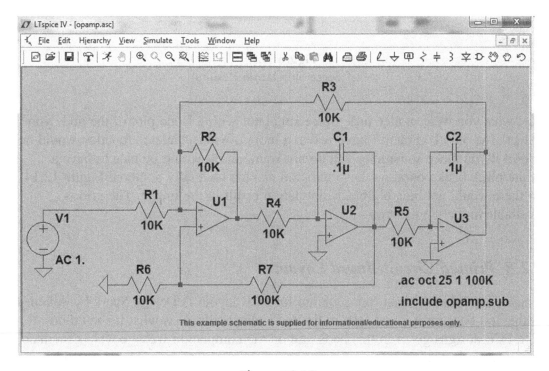

Figure 12.13
LT Spice Filter Design

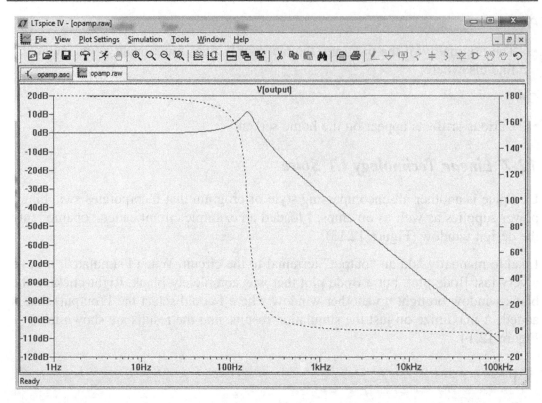

Figure 12.14
LT Spice Simulation Results

Just what you want, a nice little Spice simulator with a Bode plot of the node you select. The only comment I have is that a more obvious "delete" function would be useful if you place something you do not want, and it would be nice to have a white/black background on the simulation results. I actually doctored Figure 12.14 so there would not be a big black simulation result in the image. The grid is available in the view menu.

12.8 Printed Circuit Board Layout

I have found that the best free program for PCB layout is Design Spark by Allied. It is the best because it is not feature limited and actually does what it should do. The PCB designs given in this book and on the companion website were generated with Design Spark. Of course it is not as powerful as Cadence, Mentor, or other commercial products, but changing small prototype circuits to quite large ones can all be done quite quickly and accurately. There is even an autorouter.

Its schematic capture counterpart is a bit strange. I have not found a really good free schematic capture program yet, but the advantage of using the one from Design Spark is that it will integrate nicely with the PCB layout. That alone makes it quite useful.

12.9 Conclusions

I hope that these applications are of use to you. I have tried my best to be honest and fair. Each of them has a place on my desktop (if downloadable) and I use them occasionally. They are usable when I actually need a low/high pass with a response other than Butterworth, or a gain different from 1, or both. The very best design utilities are sometimes found on non-semiconductor sites, those done by private hobbyists. They change so quickly that it is not worth posting them here, but look around a bit. You will be surprised at what you find!

CHAPTER 13

Common Application Mistakes

13.1 Introduction

When one works as an analog applications engineer for many years supporting customer inquiries, some patterns begin to emerge. Inquiries run the gamut from those from inexperienced designers to those from analog design experts who have encountered something new and unusual that challenges even the best support engineer. Unfortunately, there are also inquiries that are bound to elicit a groan: mistakes the author has seen many times before and unfortunately will see many times again. It is hoped that this chapter will educate you about some of the most common mistakes, and save you from making them.

13.2 Op Amp Operated at Less Than Unity (or Specified) Gain

Does the phrase "unity gain stable" ring a bell? In Chapter 1, the statement was made that an op amp is least stable at its lowest specified gain. It is hoped that the engineer has taken that to heart by now! During the book tour for the first edition of *Op Amps for Everyone*, a customer approached me with a problem: they had a programmable gain op amp circuit where resistors were switched to program a gain of 1, 1/10, and 1/100 (Figure 13.1). The unity gain case worked as expected, but ringing occurred with a gain of 1/10, and sustained oscillation at a gain of 1/100. In no way do I demean the individual who made this mistake; I have been known to grab an amplifier out of a bin to construct a quick circuit, only to have it oscillate uncontrollably. Invariably, a quick glance at the data sheet reveals it is a "gain of ten" stable op amp, not unity gain stable. There is no other option at that point but to use a different op amp.

Fortunately, the solution to the problem is exceptionally easy. A voltage divider can be applied to the input of a non-inverting op amp buffer as shown in Figure 13.2.

As far as the op amp is concerned, it is operating at unity gain and is stable. The voltage divider rule is employed to calculate the correct degree of attenuation. The high input impedance of the non-inverting op amp input will not affect the voltage divider to any degree unless extremely-large-value resistors are used. If the signal

DOI: http://dx.doi.org/10.1016/B978-0-12-391495-8.00013-1

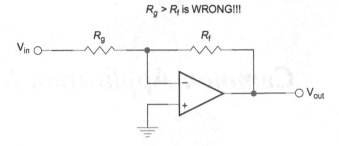

Figure 13.1
Op Amp Attenuator Done Wrong

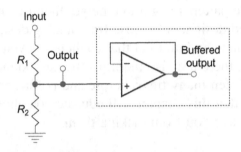

Figure 13.2
Op Amp Attenuator Done Correctly

must be inverted, then the inverting attenuator of Figure 3.3, Case 13, in Chapter 3 can be used.

13.3 Op Amp Used as a Comparator

This misapplication usually occurs in cost-sensitive pieces of equipment when a comparator is needed and a quad op amp has an unused section. I first encountered it when I discovered that the expensive telephone answering machine I had purchased quit working. Why, I asked myself, is there an open-loop op amp circuit on quarter of an LM324, and why is it interfaced to a digital logic gate? The answer was, somebody looked at the schematic symbol of an op amp and the schematic symbol of a comparator, saw that they look alike, and decided that they both work the same way!

Unfortunately, not even the internal schematic of the parts gives much indication of what is going on (Figures 13.4 and 13.5).

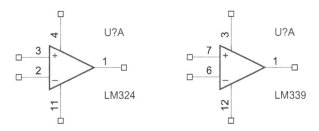

Figure 13.3
Similar Schematic Symbols, Very Different Parts!

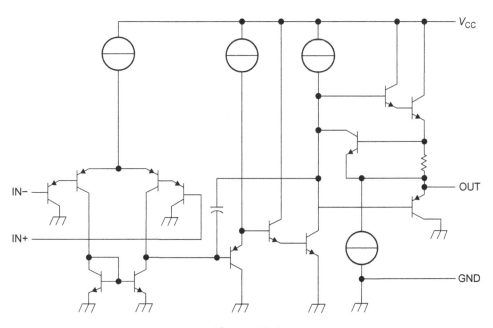

Figure 13.4
Example Op Amp Schematic

The input stages look almost identical, except the inputs are labeled opposite. The output stage of the op amp is a bit more complex, which should be a clue that something is different. The output stage of the comparator is obviously different, in that it is a single open collector. But be careful: many newer comparators have bipolar stages that are very similar in appearance to op amp output stages. An op amp has an output stage that is optimized for linear operation, while the output stage of a comparator is optimized for saturated operation.

More background about comparators versus op amps is provided in the following sections.

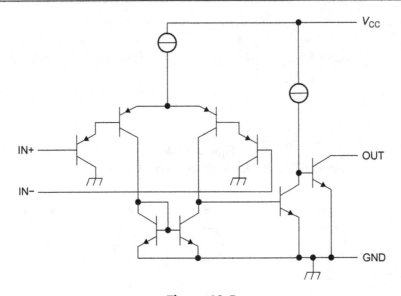

Figure 13.5
Example Comparator Schematic

13.3.1 The Comparator

A comparator is a one-bit analog-to-digital converter. It has a differential analog input and a digital output. Very few designers make the mistake of using a comparator as an op amp, because most comparators have open collector output. The output transistor of open collector comparators is characterized by low V_{CE} for switching heavy loads. The open collector structure depends on external circuitry to make the connection to power and complete the circuit. Some comparators also bring out the emitter pin, relying on the designer to complete the circuit by making both collector and emitter connections. Other comparators substitute a field-effect transistor (FET), having open drain outputs instead of an open collector. The emphasis is on driving heavy loads.

The comparator is an open-loop device, utilizing no feedback resistors. When applying a comparator, the designer compares the voltage level at two inputs. The comparator produces a digital output that corresponds to the inputs:

•   If the voltage on the non-inverting (+) input is greater than the voltage on the inverting (−) input, the output of the comparator goes to low-impedance "on" for open collector/drain outputs, and "high" for totem pole outputs.

•   If the voltage on the non-inverting (+) input is less than the voltage on the inverting (−) input, the output of the comparator goes to high-impedance "off" for open collector/drain outputs, and "low" for totem pole outputs.

13.3.2 The Op Amp

An op amp is an analog component with a differential analog input and an analog output. If an op amp is operated open loop, the output seems to act like a comparator output, but is this a good thing to do?

An op amp, being intended for closed-loop operation, is optimized for closed-loop applications. The results when an op amp is used open loop are unpredictable. No semiconductor manufacturer can or will guarantee the operation of an op amp used in an open-loop application. The analog output transistors used in op amps are designed for the output of analog waveforms, and therefore have large linear regions. The transistors will spend an inordinate amount of time in the linear region before saturation, making the rise and fall times lengthy.

In some cases, a designer may get away with using an op amp as a comparator. When an LM324 is operated in this fashion, it hits a rail and stays there, but nothing "bad" happens. The situation can change dramatically, however, when another device is substituted.

The design of an op amp output stage is bad news for a designer who needs a comparator with fast response time. The transistors used for op amp output stages are not switching transistors. They are linear devices, designed to output an accurate representation of analog waveforms. When saturated, they not only may consume more power than expected, but may also latch up. Recovery time may be very unpredictable. One batch of devices may recover in microseconds, another batch in tens of milliseconds. Recovery time is not specified, because it cannot be tested. Depending on the device, it may not recover at all. Runaway destruction of the output transistors is a distinct possibility in some rail-to-rail devices. Even the best designer may produce a saturated or even an open-loop op amp circuit without realizing it.

Oh, and the reason why my answering machine failed? The V_{OL} rail of the open-loop op amp circuit they created was above the logic threshold of the digital gate to which it was interfaced. The two levels were very close, and the slightest drift upward of the op amp output stage caused the low logic threshold never to be reached. V_{OL} is yet another op amp specification that will never be specified under open-loop conditions.

13.4 Improper Termination of Unused Sections

One of the easiest ways to unintentionally misapply an op amp is to misapply unused sections of a multiple-section integrated circuit (IC). Figure 13.6 shows the most common ways designers connect unused sections.

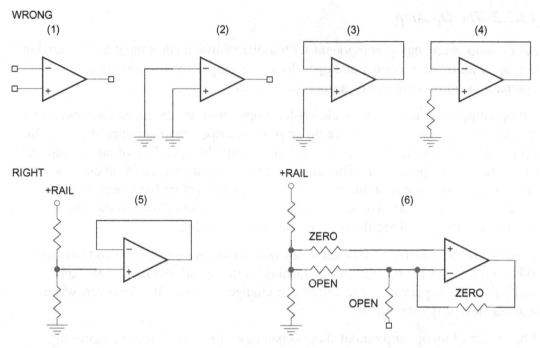

Figure 13.6
Different Ways of Dealing with Unused Op Amp Sections

Many designers know how to properly terminate unused digital inputs, hooking them to the supply or ground. These designers may not have a clue how to terminate unused op amps. Figure 13.6 demonstrates techniques I have seen, and I have given them numbers. The first four are wrong, the last two are correct:

1. This is a common mistake. Designers will assume that an op amp is like an audio amp at home, and just leave unused inputs unconnected. This is the *worst* thing that can be done to an op amp. An open-loop op amp will saturate to one voltage rail or the other. Because the inputs are floating and picking up noise, the output of the op amp will switch from rail to rail, sometimes at unpredictably high frequencies.
2. This is another really bad thing that designers occasionally do. Usually one op amp input will be slightly higher than the other owing to ground plane gradients, and the best possible scenario is that the op amp will saturate at one rail or the other. There is no guarantee it will stay there, as a slight change on one pin could cause it to switch to the other rail.
3. This is a little better than the previous case, but not that much. If the op amp is operated off a single supply, all the designer has accomplished is to ensure that the op amp will hit a rail and stay there. This can wreak havoc: self-heating,

increased power consumption. This configuration is only acceptable if the op amp is operating off split supplies.

4. Designers who are designing a board for in-circuit testing commonly do this. It still makes the op amp hit a rail if it is operated off a single rail.

5. This is the minimum recommended circuit configuration. The non-inverting input is tied to a potential between the positive and negative rail, or to ground in a split-supply system. Virtual ground may already exist in the system, making the resistors unnecessary. The op amp output will also be at virtual ground (or ground in a split-supply system).

6. This is a very good design. The designer has anticipated the possibility of system changes in the future, and designed the board so that the unused op amp section could be used by changing resistors and jumpering. The schematic shows how the unused section could be used for either an inverting or a non-inverting stage.

13.5 DC Gain

Another way designers create problems is when they forget about DC components on AC signals. Figure 13.7 illustrates this problem. When an AC signal source has a DC offset, a coupling capacitor isolates the potential in the top circuit. The DC component is rejected, and output voltage is 1 V_{AC}. If the coupling capacitor is omitted, the circuit attempts a gain of -10 on both the AC and DC components, which would be 1 V_{AC}, -50 V_{DC}. Because the power supply of the circuit limits the DC output to $+$ and -15 V_{DC}, the output will be saturated at -15 V_{DC} (minus the voltage rail limitation of the op amp).

13.6 *Current Feedback Amplifier Mistakes*

Current feedback op amps can lure you into a false sense of familiarity. After all, the familiar inverting and non-inverting gain equations are unchanged, so they are just a minor variation that works exactly the same under all circumstances. Right?

Nothing could be further from the truth — refer back to the discussion in Section 4.3 (Chapter 4) to better understand the differences.

13.6.1 *Shorted Feedback Resistor*

By far the most common mistake with current feedback amplifiers (CFAs) is that a designer will short the output directly to the inverting input (Figure 13.8).

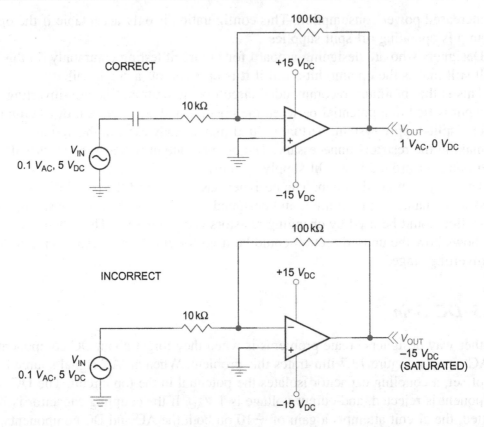

Figure 13.7
Unexpected DC Gain

The designer is invariably trying to take advantage of the speed and bandwidth of the CFA to make a unity gain buffer. Shorting the output pin to the inverting input is always a bad idea, because it will make the CFA unstable. The stability criterion for a CFA is different from that of a voltage feedback amplifier (VFA). VFA stability criterion:

$$A\beta = \frac{aR_g}{R_f + R_g}$$

CFA stability criterion:

$$A\beta = \frac{Z}{R_f\left(1 + \frac{Z_B}{R_f \| R_g}\right)}$$

As you can see, VFA stability depends on both R_f and R_g equally. But CFA stability is much more dependent on R_f. In fact, if R_f is zero, the denominator goes to zero

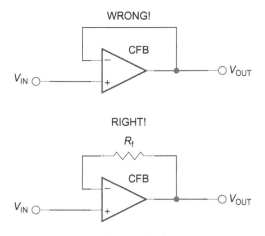

Figure 13.8
Incorrect and Correct Application of Current Feedback Amplifiers

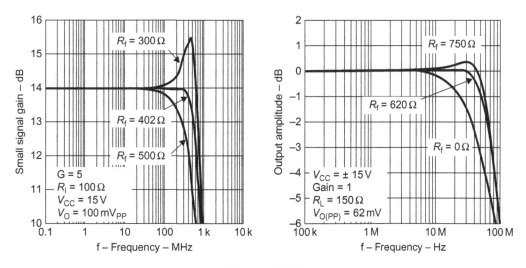

Figure 13.9
Voltage Feedback vs. Current Feedback Amplifier Stability vs. Load Resistor

and the stability criterion fails. Figure 13.9 shows this graphically in actual data sheet plots.

The effect of changing R_f only slightly has an enormous effect on the CFA response on the left, with an alarming trend as the resistor is lowered. The effect is much smaller and opposite with the VFA plot on the right. The bottom line is: stick with the recommended value of feedback resistor for current feedback op amps. This also leads to a very easy solution when a non-inverting buffer is desired: just put

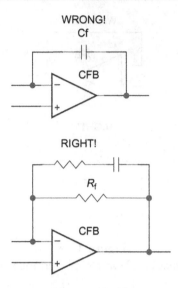

Figure 13.10
Capacitor in Feedback Loop of Current Feedback Amplifier

the recommended value of feedback resistor between the output and inverting input, and the stage will work perfectly!

13.6.2 Capacitor in the Feedback Loop

This often occurs when the designer is attempting to do active filter design with CFAs (Figure 13.10). There are very few filter topologies that will work with CFAs. Sallen–Key is one, if the proper value of feedback resistor is employed. The bottom line is that CFAs are not the best choice for active filter designs. Choose something else wherever possible.

13.7 Fully Differential Amplifier Mistakes

The reintroduction of fully differential op amps, with the new voltage output common mode (V_{OCM}) input, has made the job of interfacing to fully differential op amps and balanced lines much easier. But it has also confused a lot of designers, who have misapplied the component in subtle ways.

13.7.1 Incorrect DC Operating Point

Single-supply operation of a fully differential amplifier is very easy to mess up (Figure 13.11). What has happened here is that the two outputs have a 3.3 V DC difference between their operating point. Remember that in differential input

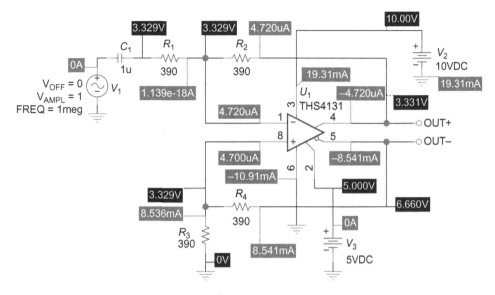

Figure 13.11
Incorrect DC Operating Point

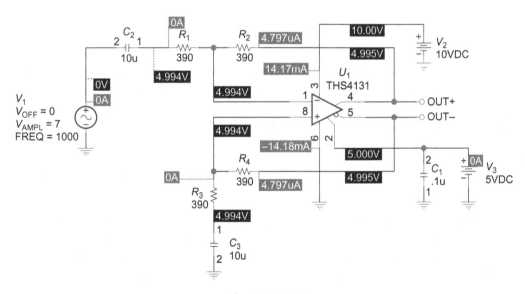

Figure 13.12
Correct DC Operating Point

circuits, there are two potential sources of DC. In this case, the designer correctly put an AC coupling capacitor between V_1 and R_1, but forgot to put one between R_3 and ground. When the second AC coupling capacitor is installed (see Figure 13.12) the correct DC operating point is established.

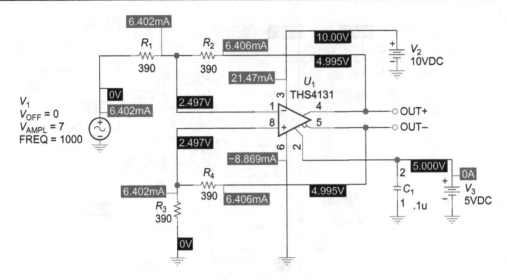

Figure 13.13
Common-Mode Error

13.7.2 Incorrect Common-Mode Range

A very subtle, but equally destructive problem often arises from incorrect application of the V_{OCM} input of the fully differential amplifier when the amplifier frequency response has to include DC, making AC coupling capacitors impossible. Consider the circuit shown in Figure 13.13.

The DC operating point appears to be correctly established, and the outputs will swing around the V_{OCM} common-mode point, which is established at 5 V by V_3. But when an AC simulation is done, the results are terrible. What happened?

The problem comes when the input voltage range does not include the negative rail, in this case ground. There are two solutions for the problem. One is to offset the inputs to the same DC level as V_{OCM}. The other is to choose a fully differential amplifier which includes the negative rail in its common-mode range. Figure 13.14 illustrates the effect of V_{OCM} on the output signals.

V_{OCM} causes problems when it forces the outputs of the amplifier too close to the power supply rails. It is best to operate both inputs and V_{OCM} as close as possible to the same DC potential.

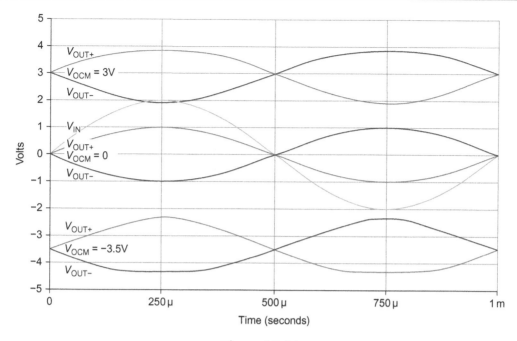

Figure 13.14
Effects of V_{OCM} on Outputs

13.7.3 Incorrect Single-Ended Termination

One of the most common applications for a fully differential op amp is single-ended to fully differential conversion. However, when the input signal must be terminated, the situation gets very complicated.

Looking at the circuit in Figure 13.15 and the equations that govern it, R_1 and R_t are cross-defined. Solving this equation for the correct values requires a goal-seeking algorithm. If the values are calculated incorrectly, the results can be:

- wrong gain
- differential offset
- unmatched differential gain
- incorrect matching impedance.

Fortunately for you, Texas Instruments has provided a calculator on their website to do this task automatically.

You might want to consider the simpler design alternative in Figure 13.16.

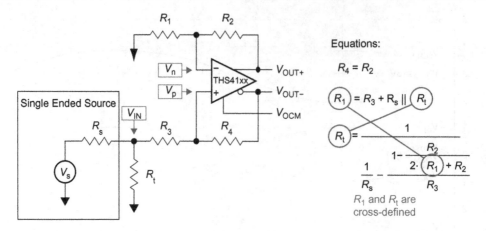

Figure 13.15
Terminating a Fully Differential Amplifier

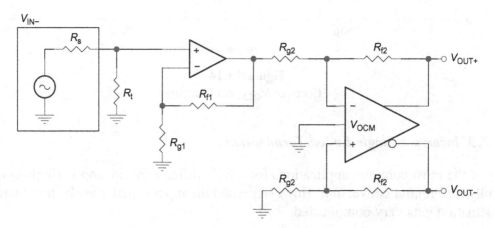

Figure 13.16
Using an Input Stage with a Fully Differential Amplifier

13.8 Improper Decoupling

In Chapter 1, I mentioned some part numbers that are etched in the memory of every design engineer, at least those involved in analog design. There is one other:

0.1 μF

Need to decouple? OK, everybody knows you put a 0.1 μF capacitor on every power supply input and the job is done, right? I can disprove that truism very easily with two words:

Cell Phone

Put your cell phone near your prototype circuit, which is bypassed with 0.1 μF, and make a call while monitoring the output on a high-bandwidth oscilloscope. You will see horrendous 2.4 GHz leakage! If your system has a speaker, and you are using a GSM (Global System for Mobile Communications) phone, at regular intervals you will hear an annoying pattern of buzz.

An alternative version of this problem came from some cellular telephone base station installers, who called in a panic: "We have 90 MHz noise running all over our system — and cannot figure out where it is coming from". A suspicion on my part asked them to tell me the exact coordinates where they were installing the system, and they provided the exact latitude and longitude. A quick check of the Federal Communications Commission (FCC) database revealed the problem. I asked them whether they were anywhere near the tower for W____ 90.5 FM, a 100,000 W NPR station listed at those coordinates. They told me on the phone that they could see the transmitter 5 feet away, they were co-locating with the station!

The point of this is that their board was bypassed with 0.1 μF capacitors. While that worked fine for the digital portions of the board, the analog portions were being clobbered by radiation from the powerful 90.5 MHz FM station. Conventional thinking is that the lower the value of capacitance, the lower the frequencies it will filter. So, 0.1 μF should get rid of just about everything because it is a very large value (relatively speaking). This conventional wisdom is wrong; the actual case is the exact opposite.

Where did the value 0.1 μF come from, anyway? A store near me used to have antique computer boards as a wall decoration. Backlit with white light, the translucent green boards made a pretty sight. But they were also populated with 0.1 μF decoupling capacitors. A quick survey of the circuitry revealed that the clock rate of the old computer had been 1 MHz.

So the 0.1 μF capacitor value seems to have come from bypassing transistor–transistor logic (TTL) in the 1960s! Isn't it time to rethink the issue a bit, in light of op amps and other analog components that can operate to frequencies of 3 GHz, especially when virtually every engineer carries a 2 W 2.4 GHz transmitter into the lab (cell phone)?

The reality of the situation is that a good 0.1 μF capacitor with an X7R dielectric exhibits a resonance in the 10 MHz region. This is due to parasitic inductance creating an LC circuit. Below 10 MHz, its impedance is capacitative, decreasing almost linearly on a logarithmic plot until it reaches the resonant frequency. Above the resonant frequency, the impedance is inductive. Since inductors resist the flow of high frequencies and only pass low frequencies, the decoupling capacitor is useless above its resonant frequency.

Looking at representative plots from capacitor manufacturers, at 100 MHz, the venerable 0.1 μF bypass capacitor has become an inductor with an inductive reactance (X_L) of at least 1 Ω. By 2.4 GHz, X_L has risen to above 10 Ω.

A good rule of thumb for effective bypassing is to put several capacitors in parallel. The standard 0.1 μF capacitor will do quite nicely for frequencies up to 10 MHz, 1000 pF NPO dielectric will do nicely up to 100 MHz, and 33 pF NPO will eliminate frequencies in the 2.4 GHz region. Bulk decoupling of the power supply as it enters the board will eliminate low-frequency ripple.

Here is a truism to replace the older one: when poor decoupling is suspected, decrease (do not increase) the value of the capacitance.

13.9 Conclusions

This chapter has been a collection of stories about mistakes made by designers over the years. My goal is not to denigrate anybody's design experience; it is to point out these commonly made errors so you will not make them. Engineering experience is often a memory of failures and what the solutions were, and learning how not to make the same mistakes again. Any time you have an opportunity to learn from other engineers' experience, learn from them with a grateful heart.

I have received a lot of criticism for items in this chapter, particularly in Sections 13.2 and 13.3. A lot of engineers have done these things for years, and feel threatened when I point out that it is incorrect. I do not have a lot of respect for the old "I've done it for years and it works" argument. I would counter with "You have been very lucky not to have been bitten by it".

Just because something works does not automatically make it good design practice. It may be very marginal, on the edge of failure, and you will not notice it until products fail in use, causing expensive and embarrassing returns. This had led to many a career being wrecked, especially if the recall/return affects a large production run.

Correct the obvious problems before they have a chance to mess up your product. Do not reinvent somebody's past mistakes. You will invent plenty of your own mistakes along the way without repeating somebody else's.

Appendix A: Understanding Op Amp Parameters

A.1 Introduction

This appendix explains op amp data sheet parameters. By its very nature, it will be long and technical, which is why it is included as an appendix instead of a chapter. Although I have covered most of the common parameters in the chapters of the book, there is some detail here that you should know if you need to maximize the performance of your op amp stage.

Parameters are listed in alphabetical order, not order of importance, because different parameters are important for different applications. I have included "relevance" with each parameter (or group of parameters) in its description, which will allow you to quickly search for those parameters more important for your application. I note a handful of parameters that are important to a wide range of applications, ones you should always be aware of no matter what you are designing. They are **boldfaced** in Table A.1 and noted in their descriptive sections.

Units listed in the "units" column of Table A.1 are standard SI units of measure. The base unit is specified, not multiplier prefixes such as "μ" (micro). Multiplier prefixes, however, are used in data sheets.

There are usually three main sections of electrical tables in op amp data sheets.

A.1.1 Absolute Maximum Ratings

Absolute maximum ratings are those limits beyond which the life of individual devices may be impaired and are never to be exceeded in service or testing. Limits, by definition, are maximum ratings, so if double-ended limits are specified, the term will be defined as a range (e.g. operating temperature range).

A.1.2 Recommended Operating Conditions

Recommended operating conditions have a similarity to maximum ratings in that operation outside the stated limits could cause unsatisfactory performance. Recommended operating conditions, however, do not carry the implication of device damage if they are exceeded.

Table A.1: Op Amp Condition and Parameter Table

Abbreviation	Parameter	Units
αI_{IO}	Temperature coefficient of input offset current	A/°C
αV_{IO} or α_{VIO}	Temperature coefficient of input offset voltage	V/°C
A_D	Differential gain error	%
A_m	Gain margin	dB
A_{OL}	**Open-loop voltage gain**	**dB**
A_V	Large-signal voltage amplification (gain)	dB
A_{VD}	Differential large-signal voltage amplification	dB
B_1	**Unity gain bandwidth**	**Hz**
B_{OM}	Maximum-output-swing bandwidth	Hz
BW	Bandwidth	Hz
c_i	Input capacitance	F
C_{ic} or $C_{i(c)}$	Common-mode input capacitance	F
C_{id}	Differential input capacitance	F
C_L	Load capacitance	F
$\Delta V_{DD\pm}$ (or CC ±)/ ΔV_{IO}, or k_{SVS}	Supply voltage sensitivity	dB
CMRR or k_{CMR}	**Common-mode rejection ratio**	**dB**
f	Frequency	Hz
GBW	**Gain bandwidth product**	**Hz**
$I_{CC-\ (SHDN)}$, $I_{DD-\ (SHDN)}$	Supply current (shutdown)	A
I_{CC}, I_{DD}	**Supply current**	**A**
I_I	Input current range	A
I_{IB}	**Input bias current**	**A**
I_{IO}	Input offset current	A
I_n	Input noise current	A/$\sqrt{Hz}$
I_O	Output current	A
I_{OL}	Low-level output current	A
I_{OS}, or I_{SC}	Short-circuit output current	A
k_{CMR} or CMRR	Common-mode rejection ratio	dB
k_{SVR}	**Supply rejection ratio**	**dB**
k_{SVS}	**Supply voltage sensitivity**	**dB**
P_D	Power dissipation	W
PSRR	**Power supply rejection ratio**	**dB**
θ_{JA}	Junction to ambient thermal resistance	°C/W
θ_{JC}	Junction to case thermal resistance	°C/W
r_i	Input resistance	Ω
r_{id}, $r_{i(d)}$	Differential input resistance	Ω
R_L	Load resistance	Ω
R_{null}	Null resistance	Ω
r_o	Output resistance	Ω
R_S	Signal source resistance	Ω
R_t	Open-loop transresistance	Ω
SR	**Slew rate**	**V/s**
T_A	Operating temperature	°C
t_{DIS} or $t_{(off)}$	Turn-off time (shutdown)	s
t_{EN} or $t_{(on)}$	Turn-on time (shutdown)	s
t_f	Fall time	s
THD	Total harmonic distortion	%
THD + N	Total harmonic distortion plus noise	%

(Continued)

Table A.1: (Continued)

Abbreviation	Parameter	Units
T_J	Maximum junction temperature	°C
t_r	Rise time	s
t_s	Settling time	s
T_S or T_{stg}	Storage temperature	°C
V_{CC}, V_{DD}	**Supply voltage**	**V**
V_I	**Input voltage range**	**V**
V_{ic}	Common-mode input voltage	V
V_{ICR}	Input common-mode voltage range	V
V_{ID}	Differential input voltage	V
V_{DIR}	Differential input voltage range	V
$V_{IH\text{-}SHDN}$ or $V_{(ON)}$	Turn-on voltage (shutdown)	V
$V_{IL\text{-}SHDN}$ or $V_{(OFF)}$	Turn-off voltage (shutdown)	V
V_{IN}	Input voltage (DC)	V
V_{IO}, V_{OS}	**Input offset voltage**	**V**
V_n	**Equivalent input noise voltage**	**V/√Hz**
$V_{N(PP)}$	Broad band noise	V P–P
V_{OH}	**High-level output voltage**	**V**
V_{OL}	**Low-level output voltage**	**V**
$V_{OM\pm}$	Maximum peak-to-peak output voltage swing	V
$V_{O(PP)}$	Peak-to-peak output voltage swing	V
$V_{(STEP)PP}$	Step voltage peak-to-peak	V
X_T	Crosstalk	dB
Z_o	Output impedance	Ω
Z_t	Open-loop transimpedance	Ω
Φ_D	Differential phase error	°
Φ_m	**Phase margin**	°
	Bandwidth for 0.1 dB flatness	Hz
	Case temperature for 60 seconds °C	°C
	Continuous total dissipation	W
	Differential gain error	%
	Differential phase error	°
	Duration of short-circuit current	s
	Input offset voltage long-term drift	V/month
	Lead temperature for 10 or 60 seconds	°C

A.1.3 Electrical Characteristics

Electrical characteristics are measurable electrical properties of a device inherent in its design. They are used to predict the performance of the device as an element of an electrical circuit. The measurements that appear in the electrical characteristics tables are based on the device being operated within the recommended operating conditions.

Some of Table A.1 is composed of parameters and some test conditions. Test conditions are conditions placed on the op amp when the parameters are measured. Some abbreviations are used for both a condition and a parameter.

A.2 Temperature Coefficient of the Input Offset Current (αI_{IO})

αI_{IO} specifies the expected input offset current drift over temperature. Its units are $\mu A/°C$. I_{IO} is measured at the temperature extremes of the part, and αI_{IO} is computed as $\Delta I_{IO}/\Delta °C$.

Relevance: DC applications with current input, when operated in an environment where the temperature can vary.

A.3 Temperature Coefficient of the Input Offset Voltage (αV_{IO} or α_{VIO})

αV_{IO} specifies the expected input offset drift over temperature. Its units are $V/°C$. V_{IO} is measured at the temperature extremes of the part, and αV_{IO} is computed as $\Delta V_{IO}/\Delta °C$.

Relevance: DC applications with voltage input, when operated in an environment where the temperature can vary.

A.4 Differential Gain Error (A_D)

The differential gain error parameter, A_D, is defined as the change in AC gain with change in DC level. The AC signal is 40 IRE (0.28 V PK) and the DC level change is ± 100 IRE (± 0.7 V). Typically tested at 3.58 MHz (NTSC) or 4.43 MHz (PAL) carrier frequencies. It is represented in units of %.

Relevance: With the conversion to digital broadcast video, this parameter is quickly becoming less relevant.

A.5 Gain Margin Parameter (A_m)

Gain margin, A_m, is defined as the absolute value of the difference in gain between the unity gain point and the gain at the $-180°$ phase shift point. It is measured open loop and expressed in units of decibels, dB.

Gain margin (A_m) and phase margin (Φ_m) are different ways of specifying the stability of the circuit. Figure A.1 shows the gain margin (and companion parameter phase margin) graphically.

Relevance: While this is a close cousin to the extremely important phase margin (Section A.72), this parameter is seldom used.

A.6 Open-Loop Voltage Gain Parameter (A_{OL})

Important

Figure A.2 will be used for several parameters.

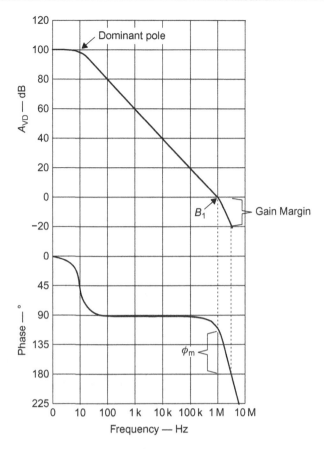

Figure A.1
Gain and Phase Margin

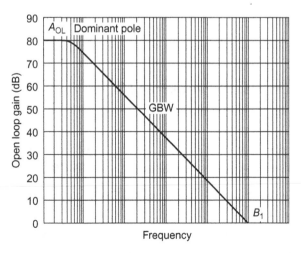

Figure A.2
Open-Loop Parameters

The open-loop voltage gain parameter, A_{OL}, is defined as the ratio of change in output voltage to the change in voltage across the input terminals. It is expressed either unitless, V/V, or more commonly in dB. When expressed as a single value, it refers to the portion of the upper graph in Figure A.2 that is horizontal, so the single value is valid only for DC and low frequencies. Once the frequency has risen to the point that the dominant pole is reached, A_{OL} has a frequency dependence and is replaced by the gain bandwidth product (GBW) described in Section A.19.

Relevance: This parameter is important in high-gain DC and low-frequency applications, because it can affect the gain of the circuit.

A.7 Large-Signal Voltage Amplification Gain Condition (A_V)

The large-signal voltage amplification or gain condition, A_V, is defined as the ratio of change in output voltage to the change in voltage across the input terminals that is set up for a test of parameters such as Z_O or THD + N. It is expressed either unitless or in dB.

Relevance: This condition is seldom used except in conjunction with the above-mentioned parameters.

A.8 Differential Large-Signal Voltage Amplification Parameter (A_{VD})

The differential large-signal voltage amplification parameter, A_{VD}, is defined as the ratio of change in output voltage to the change in voltage across the input terminals. It is expressed either unitless or in dB. A_{VD} is sometimes referred to as differential voltage gain. A_{VD} is similar to the open-loop gain A_{OL} of the amplifier, except A_{OL} is usually measured without any load. A_{VD} is usually measured with a load. Both parameters are measured open loop.

Relevance: This parameter is seldom used.

A.9 Unity Gain Bandwidth Parameter (B_1)

Important

Unity gain bandwidth, B_1, is defined as the range of frequencies within which the open-loop voltage amplification is greater than or equal to unity (0 dB). B_1 is expressed in units of Hz.

Relevance: This parameter is a single point on the GBW plot, where it crosses the 0 dB (or the specified minimum gain) axis. You often see it prominently displayed as a marketing bullet item for high-speed op amps, but do not be fooled into designing with it alone! While it is important, it is the whole open-loop response curve that you need because you will almost never design an op amp circuit at its frequency limit. A_{OL} and B_1 can be used to approximate the open-loop response for voltage feedback amplifiers, knowing that a line passing through B_1 at -20 dB per decade will intersect the A_{OL} horizontal line at the position of the dominant pole (see Figure A.2)

A.10 Maximum-Output-Swing Bandwidth Parameter (B_{OM})

The maximum-output-swing bandwidth parameter, B_{OM}, is defined as the maximum frequency which the output swing is above a specified value or at the extents of its linear range. B_{OM} is also called full power bandwidth. B_{OM} is expressed in units of Hz.

The limiting factor for B_{OM} is slew rate (SR). As the frequency gets higher and higher the output becomes slew rate limited and cannot respond quickly enough to maintain the specified output voltage swing. Equation A.1 expresses the relationship between B_{OM} and SR:

$$B_{OM} = \frac{SR}{2\pi V_{(PP)}} \tag{A.1}$$

Relevance: This parameter is related to the slew rate (Section A.41), which is the more important parameter. This parameter is primarily important when the op amp has a large voltage swing on a non-sinusoidal signal source.

A.11 Bandwidth Condition (BW)

Bandwidth, BW, is defined as a maximum frequency minus a minimum frequency. It is included here for completeness. It is used when describing op amp parameters such as small signal (-3 dB), 0.1 dB flatness, and full power. BW is expressed in units of Hz.

Relevance: This condition is important when given in the context of another parameter. Bandwidth with no other information is in need of the conditions that make it a parameter.

A.12 Input Capacitance Parameter (C_i)

The input capacitance parameter, C_i, is defined as the capacitance between the input terminals of an op amp with either input grounded. It is expressed in units of F.

C_i is one of a group of parasitic elements affecting input impedance. Figure A.2 shows a model of the resistance and capacitance between each input terminal and ground and between the two terminals. There is also parasitic inductance, but the effects are negligible at low frequency. Input impedance is a design issue when the source impedance is high. The input loads the source.

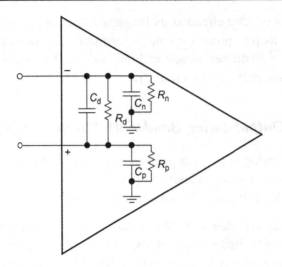

Figure A.3
Input Parasitic Elements

Input capacitance, C_i, is measured between the input terminals with either input grounded. C_i is usually a few pF. In Figure A.3, if V_p is grounded, then $C_i = C_d || C_n$.

Sometimes common-mode input capacitance, C_{ic}, is specified. In Figure A.3, if V_p is shorted to V_n, then $C_{ic} = C_p || C_n$. C_{ic} is the input capacitance a common-mode source would see referenced to ground.

Relevance: This and the following two parameters are of most relevance to high-speed applications, particularly those involving step responses. That can make them fairly important, but since they are rolled into the slew rate (SR, Section A.41), I do not include them on the "important" list. If you are tracking them separately, you have a fairly advanced application that is beyond the scope of this book.

A.13 Common-Mode Input Capacitance Parameter (C_{ic} or $C_{i(c)}$)

The common-mode input capacitance parameter, C_{ic} or $C_{i(c)}$, is defined as the input capacitance a common-mode source would see to ground. It is expressed in units of F.

C_{ic} is one of a group of parasitic elements affecting input impedance. Figure A.2 shows a model of the resistance and capacitance between each input terminal and ground and between the two terminals. There is also parasitic inductance, but the effects are negligible at low frequency. Input impedance is a design issue when the source impedance is high. The input loads the source.

Input capacitance, C_i, is measured between the input terminals with either input grounded. C_i is usually a few pF. In Figure A.3, if V_p is grounded, then $C_i = C_d||C_n$.

Sometimes common-mode input capacitance, C_{ic}, is specified. In Figure A.3, if V_p is shorted to V_n, then $C_{ic} = C_p||C_n$. C_{ic} is the input capacitance a common-mode source would see when referenced to ground.

A.14 Differential Input Capacitance Parameter (C_{id})

The differential input capacitance parameter, C_{id}, is the same as the common-mode input capacitance, C_{ic}. It is the input capacitance a common-mode source would see to ground. It is expressed in units of F.

A.15 Load Capacitance Condition (C_L)

The load capacitance condition, C_L, is defined as the capacitance between the output terminal of an op amp and ground. It is expressed in units of F.

C_L is a capacitative load that is sometimes connected to an op amp when parameters such as SR, t_s, Φ_m, or A_m are being tested.

Relevance: You need to keep in mind that load capacitance affects op amp stability: some op amps are made to drive capacitive loads and some are not. If you need to drive a capacitative load, select an op amp made for the job. A lot of designers get into trouble, for example, when they generate a reference voltage with an op amp, and then treat it like a voltage regulator and hang a large capacitor across the output. Bad idea! Use a series resistance before the capacitor.

A.16 Supply Voltage Sensitivity ($\Delta V_{DD\pm(or\ CC\pm)}/\Delta V_{IO}$, or k_{SVR}, or PSRR)

Important

The power supply rejection ratio, $\Delta V_{DD\pm(or\ CC\pm)}/\Delta V_{IO}$, is the same as the supply rejection ratio, k_{SVR} and PSRR. It is defined as the absolute value of the ratio of the change in supply voltages to the resulting change in input offset voltage. Typically, both supply voltages are varied symmetrically. It is expressed in dB.

The power voltage affects the bias point of the input differential pair. Because of the inherent mismatches in the input circuitry, changing the bias point changes the offset voltage which, in turn, changes the output voltage.

For a dual-supply op amp, $k_{SVR} = \Delta V_{CC\pm}/\Delta V_{OS}$ or $\Delta V_{DD\pm}/\Delta V_{OS}$. The term $\Delta V_{CC\pm}$ means that the plus and minus power supplies are changed symmetrically. For a single-supply op amp, $k_{SVR} = \Delta V_{CC}/\Delta V_{OS}$ or $\Delta V_{DD}/\Delta V_{OS}$. Also note that the mechanism that produces k_{SVR} is the same as for CMRR. Therefore, k_{SVR} as published in the data sheet is a DC parameter like CMRR. When k_{SVR} is graphed vs. frequency, it falls off as the frequency increases.

Relevance: This will affect any high-gain, low-noise application. This parameter also becomes important if there are high levels of ripple on the power supply, as happens when the power supply is switching. Proper bypassing techniques must be used in any case, but particularly if the power supply is noisy.

A.17 Common-Mode Rejection Ratio Parameter (CMRR or k$_{CMR}$)

Important

The common-mode rejection ratio parameter, CMRR or k_{CMR}, is defined as the ratio of differential voltage amplification to common-mode voltage amplification. This is measured by determining the ratio of a change in input common-mode voltage to the resulting change in input offset voltage. It is expressed in dB.

Ideally, CMRR or k_{CMR} would be infinite with common-mode voltages being totally rejected.

The common-mode input voltage affects the bias point of the input differential pair. Because of the inherent mismatches in the input circuitry, changing the bias point changes the offset voltage which, in turn, changes the output voltage. The real mechanism at work is $\Delta_{VOS}/\Delta V_{COM}$.

A common source of common-mode interference voltage is 50 Hz or 60 Hz AC noise. Care must be used to ensure that the CMRR of the op amp is not degraded by other circuit components. High values of resistance make the circuit vulnerable to common-mode (and other) noise pick-up. It is usually possible to scale resistors down and capacitors up to preserve circuit response.

Relevance: Any gain stage amplifying low-level signals. This is one of those parameters that can be easily overlooked and cause poor circuit performance. Bad circuit board layout techniques can aggravate or even be the dominant source of common-mode noise. Wherever possible in low signal level conditions, use fully differential op amps to combat common-mode noise.

A.18 Frequency Condition (f)

Frequency condition, f, is the frequency under which a parameter is tested. It is expressed in Hz.

Relevance: This condition has been listed for completeness. It is relevant and self-evident for all designs.

A.19 Op Amp Gain Bandwidth Product Parameter (GBW)

Important

Gain bandwidth product, GBW, is defined as the product of the open-loop voltage gain and the frequency at which it is measured. GBW is expressed in units of Hz. Figure A.2 shows the GBW graphically: it is the portion that decreases linearly at 20 dB per decade.

GBW is constant for voltage feedback amplifiers. It does not have much meaning for current feedback amplifiers because there is not a linear relationship between gain and bandwidth.

When an op amp is selected for a specific application both the bandwidth and the slew rate should be taken into account (along with other factors including power consumption, distortion, and price).

Relevance: This parameter is used for high-speed application circuits. "High speed", however, can get pretty low if high-gain or filter circuits are involved.

A.20 Supply Current (Shutdown) Parameter ($I_{CC(SHDN)}$ or $I_{DD(SHDN)}$)

The supply current (shutdown) parameter, $I_{CC(SHDN)}$ or $I_{DD(SHDN)}$, is defined as the current into the V_{CC+} (V_{DD+}) or V_{CC-} (V_{DD-}) terminal of the amplifier while it is turned off. It is expressed in units of A.

Relevance: This is only relevant when the op amp has a shutdown function.

A.21 Supply Current Parameter (I_{CC} or I_{DD})

Important

The supply current parameter, I_{CC} or I_{DD}, is defined as the current into the V_{CC+} (V_{DD+}) or V_{CC-} (V_{DD-}) terminal of the op amp while it is operating without load and the input and/or output is at virtual ground. It is expressed in units of A.

Relevance: All op amp applications, but those that rely on battery power in particular.

A.22 Input Current Range Parameter (I_I)

The input current range parameter, I_I, is defined as the amount of current that can be sourced or sinked by the op amp input. It is usually specified as an absolute maximum rating expressed in units of A.

Relevance: Transducers that have a current output instead of voltage output.

A.23 Input Bias Current Parameter (I_{IB})

Important

The input bias current parameter, I_{IB}, is defined as the average of the currents into the two input terminals with the output at a specified level. It is expressed in units of A.

The input circuitry of all op amps requires a certain amount of bias current for proper operation. The input bias current, I_{IB}, is computed as the average of the two inputs:

$$I_{IB} = \frac{(I_N + I_P)}{2} \tag{A.2}$$

CMOS and JFET inputs offer much lower input current than standard bipolar inputs.

Input bias current is of concern when the source impedance is high. If the op amp has high input bias current, it will load the source and a lower than expected voltage is seen. If the source impedance is high, the best solution is to use an op amp with either CMOS or JFET input. The source impedance can also be lowered by using a buffer stage to drive the op amp that has high input bias current.

Relevance: I list this one as important because it is often overlooked, resulting in non-functional circuits. What usually happens is that somebody is using a non-inverting op amp gain circuit in an AC coupled configuration, and has nothing but the coupling capacitor on the op amp input. If the op amp were ideal, the non-inverting input would float to the potential of the inverting input, and the circuit would work. But an op amp is not an ideal component, and requires some kind of bias on the non-inverting input. A very easy way to do this is to equal resistors to the positive and negative rail, which also is handy for setting the DC operating point of the stage. Alternately, if a reference is available, a single large-value resistor can be connected to it. Beware, though, of high-temperature circuits. The input bias current can vary by decades. A circuit that works well at ambient may not work at all at high temperatures because there is insufficient input bias current.

A.24 Input Offset Current Parameter (I_{IO})

The input offset current parameter, I_{IO}, is defined as the difference between the currents into the two input terminals of an op amp with the output at the specified level. It is expressed in units of A.

Relevance: Transducers that have a current output instead of voltage output. If that is the case, consider this to be an important parameter, replacing the input offset voltage. Because current amplification is comparatively rare, I do not include this on the important parameters list for all users.

A.25 Input Noise Current Parameter (I_n)

The input noise current parameter, I_n, is defined as the internal noise current reflected back to an ideal current source in parallel with the input pins. It is expressed in units of $A/\sqrt{Hz}$.

It is important for a designer to calculate noise that the device will deliver in an application. The simplest way to calculate this noise is to use the following equation:

$$e_{nt} = \sqrt{V_n^2 + (I_n \times R_s)^2} \tag{A.3}$$

where e_{nt} = total noise voltage, V_n = voltage noise ($nV/\sqrt{Hz}$), I_n = current noise ($pA/\sqrt{Hz}$), and R_s = source resistance (Ω).

Relevance: Transducers that have a current output instead of voltage output. If that is the case, consider this to be an important parameter, replacing the input voltage noise. Because current amplification is comparatively rare, I do not include this on the important parameters list for all users.

A.26 Output Current Parameter (I_O)

The output current parameter, I_O, is defined as the amount of current that may be drawn from the op amp output. Usually I_O is expressed in units of A.

Relevance: This parameter is important if you are driving a heavy load. Most voltage feedback op amps can drive about 600Ω; if you load them any more this results in serious degradation of the output stage parameters, or even damage. Current feedback op amps usually have better output drive, line driver ICs even more. Do not overlook audio amplifier ICs, many of which have op amp-like input and feedback topologies, and can be used to drive a fair amount of power. There are even higher output current op amps in TO-3 packages, which can operate off hazardous voltages, so be careful!

A.27 Low-Level Output Current Condition (I_{OL})

The low-level output current condition, I_{OL}, is defined as the current into an output that is supplied during the test for V_{OL}. It is usually expressed in units of A.

Relevance: Seldom used.

A.28 Short-Circuit Output Current Parameters (I_{OS} or I_{SC})

The short-circuit output current parameter, I_{OS} or I_{SC}, is defined as the maximum output current available from the amplifier with the output shorted to ground, to either supply, or to a specified point. Sometimes a low-value series resistor is specified. It is usually expressed in units of A.

It is important to observe power dissipation ratings to keep the junction temperature below the absolute maximum rating when the output is heavily loaded or shorted. See the absolute maximum ratings section of the part's data sheet for more information.

Relevance: Seldom used except to define failure modes of a circuit.

A.29 Supply Rejection Ratio Parameter (k_{SVR})

Important

The supply rejection ratio, k_{SVR}, is the same as the power supply rejection ratio, PSRR. It is defined as the absolute value of the ratio of the change in supply voltages to the resulting change in input offset voltage. Typically, both supply voltages are varied symmetrically. It is expressed in dB.

The power voltage affects the bias point of the input differential pair. Because of the inherent mismatches in the input circuitry, changing the bias point changes the offset voltage which, in turn, changes the output voltage.

For a dual-supply op amp, $k_{SVR} = \Delta V_{CC\pm}/\Delta V_{OS}$ or $\Delta V_{DD\pm}/\Delta V_{OS}$. The term $\Delta V_{CC\pm}$ means that the plus and minus power supplies are changed symmetrically. For a single-supply op amp, $k_{SVR} = \Delta V_{CC}/\Delta V_{OS}$ or $\Delta V_{DD}/\Delta V_{OS}$. Also note that the mechanism that produces k_{SVR} is the same as for CMRR. Therefore, k_{SVR} as published in the data sheet is a DC parameter like CMRR. When k_{SVR} is graphed vs. frequency, it falls off as the frequency increases.

Relevance: Important for high gain and/or low noise systems. It is also important if the op amps are powered off a switching power supply. Proper bypassing techniques must be used.

A.30 Power Dissipation Parameter (P_D)

The power dissipation, P_D, is defined as the power supplied to the device less any power delivered from the device to a load. Note that at no load, $P_D = V_{CC+} \times ts\}I_{CC}$ or $P_D = V_{DD+} \times I_{DD}$. It is expressed in units of W.

Relevance: This parameter is important if the op amp is driving a high load current.

A.31 Power Supply Rejection Ratio Parameter (PSRR)

The power supply rejection ratio, PSRR, is the same as the supply rejection ratio, k_{SVR} (see Section A.29).

A.32 Junction to Ambient Thermal Resistance Parameter (θ_{JA})

The junction to ambient thermal resistance parameter, θ_{JA}, is defined as the ratio of the difference in temperature from the die junction to the ambient air and the power dissipated by the die. θ_{JA} is expressed in units of °C/W.

θ_{JA} is dependent on the case to the ambient thermal resistance as well as the θ_{JC} parameter. θ_{JA} is a better indicator of thermal resistance when the package is not well thermally sinked to other components in the assembly. θ_{JA} is listed in the data sheet for different packages. It is useful for evaluating which package is least likely to overheat and to determine what the die temperature is when the ambient temperature and power dissipation are known.

Relevance: This parameter is important if the op amp is driving a high load current. It is used to determine heatsinking requirements.

A.33 Junction to Case Thermal Resistance Parameter (θ_{JC})

The junction to case thermal resistance parameter, θ_{JC}, is defined as the ratio of the difference in temperature from the die junction to the case and the power dissipated by the die. θ_{JC} is expressed in units of °C/W.

θ_{JC} is not dependent on the case to the ambient thermal resistance as is the θ_{JA} parameter. θ_{JC} is a better indicator of thermal resistance when the package is to be thermally sinked to other components in the assembly.

θ_{JC} is listed in the data sheet for different packages. It is useful for evaluating which package is least likely to overheat and to determine what the die temperature is when the case temperature and power dissipation are known.

Relevance: This parameter is important if the op amp is driving a high load current. It is used to determine heatsinking requirements.

A.34 Input Resistance Parameter (r_i)

The input resistance parameter, r_i, is defined as the DC resistance between the input terminals with either input grounded. It is expressed in units of Ω.

r_i is one of a group of parasitic elements affecting input impedance. Figure A.2 shows a model of the resistance and capacitance between each input terminal and ground and between the two terminals. There is also parasitic inductance, but the effects are negligible at low frequency. Input impedance is a design issue when the source impedance is high. The input loads the source.

Input resistance, r_i, is the resistance between the input terminals with either input grounded. In Figure A.3, if V_p is grounded, then $r_i = R_d || R_n$. r_i ranges from $10^7 \Omega$ to $10^{12} \Omega$, depending on the type of input.

Sometimes common-mode input resistance, r_{ic}, is specified. In Figure A.3, if V_p is shorted to V_n, then $r_{ic} = R_p || R_n$. r_{ic} is the input resistance a common-mode source would see referenced to ground.

Relevance: This parameter is one of the parasitic elements shown in Figure A.3. As it is usually very large, it is only relevant when the input source is high resistance, such as a photomultiplier tube.

A.35 Differential Input Resistance Parameter (r_{id} or $r_{i(d)}$)

The differential input resistance, r_{id} or $r_{i(d)}$, is defined as the small-signal resistance between two ungrounded input terminals. It is expressed in units of Ω.

r_{id} is one of a group of parasitic elements affecting input impedance. Figure A.2 shows a model of the resistance and capacitance between each input terminal and ground and between the two terminals. There is also parasitic inductance, but the effects are negligible at low frequency. Input impedance is a design issue when the source impedance is high. The input loads the source.

Input resistance, r_i, is the resistance between the input terminals with either input grounded. In Figure A.3, if V_p is grounded, then $r_i = R_d || R_n$. r_i ranges from $10^7 \Omega$ to $10^{12} \Omega$, depending on the type of input.

Sometimes common-mode input resistance, r_{ic}, is specified. In Figure A.3, if V_p is shorted to V_n, then $r_{ic} = R_p || R_n$. r_{ic} is the input resistance a common-mode source would see referenced to ground. In Figure A.3 $r_{id} = R_d$.

Relevance: This parameter is one of the parasitic elements shown in Figure A.3. As it is usually very large, it is only relevant when the input source is high resistance, such as a photomultiplier tube.

A.36 Load Resistance Condition (R_L)

The load resistance condition, R_L, is defined as the DC resistance that is attached from the output of an op amp to ground during a test for a parameter such as A_{VD}, SR, THD + D, $t_{(on)}$, $t_{(off)}$, GBW, t_s, Φ_m, and A_m. It is expressed in units of Ω.

Relevance: Normal op amp stage design does not usually involve driving a heavy load. Voltage feedback amplifiers are generally capable of driving a 600Ω load without degradation of specifications, while current feedback amplifiers have heftier output stages meant to drive DSL lines at 100Ω. If you have any doubt about your load, pay attention to this condition.

A.37 Null Resistance Condition (R_L)

The null resistance condition, R_L, is defined as the DC resistance that is attached in series with C_L when testing for parameters such as phase margin and gain margin. It is expressed in units of Ω.

Relevance: Seldom used.

A.38 Output Resistance Parameter (r_o)

The output resistance parameter, r_o, is defined as the DC resistance that is placed in series with the output of an ideal amplifier and the output terminal for simulation of the real device. It is expressed in units of Ω.

Relevance: Normal op amp stage design does not use this parameter, which can be assumed to be 0Ω. However, if you plan on driving a heavier load, you need to take this resistance into account when designing for the load and especially the output voltage swing.

A.39 Signal Source Condition (R_S)

The signal source condition, R_S, is defined as the output resistance of a signal source. It is expressed in units of Ω. R_S is used as a test condition when measuring parameters such as V_{IO}, α_{VIO}, I_{IO}, I_{IB}, and CMMR. A typical value for R_S used in these parameter tests is 50Ω.

Relevance: Seldom used.

A.40 Open-Loop Transresistance Parameters (R_t)

In a transimpedance or current feedback amplifier, the open-loop transresistance parameter, R_t, is defined as the ratio of change in DC output voltage to the change in DC current at the inverting input. It is expressed in units of Ω.

Relevance: Current feedback amplifier designs.

A.41 Op Amp Slew Rate Parameter (SR)

Important

The slew rate parameter, SR, is defined as the rate of change in the output voltage caused by a step change at the input. It is expressed in V/s. The SR parameter of an op amp is the

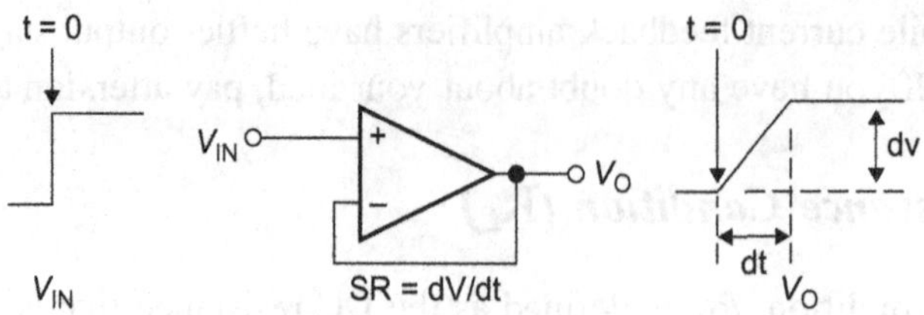

Figure A.4
Slew Rate

maximum SR it will pass and is generally specified with a gain of 1. Figure A.4 shows SR graphically.

In order for an amplifier to pass a signal without distortion due to insufficient SR, the amplifier must have at least the maximum SR of the signal. The maximum SR of a sine wave occurs as it crosses zero. The following equation defines this slew rate:

$$SR = 2\pi fV \tag{A.4}$$

where f = frequency of the signal, and V = peak voltage of the signal.

The SR is sometimes represented as $SR+$ and $SR-$. $SR+$ is the abbreviation for the slew rate for a positive transition and $SR-$ is the abbreviation for the slew rate for a negative transition. Many applications are best served when $SR+$ and $SR-$ are the same magnitude.

The primary factor controlling SR in most op amps is an internal compensation capacitor which is added to make the op amp unity gain stable. When an op amp is selected for a specific application both the bandwidth and the SR should be taken into account.

Relevance: This parameter is important when the input waveform varies rapidly and/or has large steps in response over a short period.

A.42 Operating Free-Air Temperature Condition (T_A)

The operating free-air temperature condition, T_A, is defined as the free-air temperature over which the op amp is being operated. Some of the other parameters may change with temperature, leading to degraded operation at temperature extremes. T_A is expressed in units of °C.

A range of T_A is listed in a data sheet's absolute maximum ratings table because stress beyond those listed may cause permanent damage to the device. Functional operation to this limit is not implied and might affect reliability.

Relevance: Systems which will encounter large variations of temperature.

A.43 Turn-Off Time (Shutdown) Parameter (t_{DIS} or $t_{(off)}$)

The turn-off time (shutdown) parameter, t_{DIS} or $t_{(off)}$, is defined as the time from when the turn-off voltage is applied to the shutdown pin to when the supply current has reached half of its final value. It is expressed in units of s.

You need to carefully read the data sheet to understand the logic level of the enable/disable function. Some are referenced to the most negative supply, which can force you to design a logic-level shifting circuit.

Relevance: This parameter is only relevant when the op amp has an enable/disable input, and when this disable/enable function is used.

A.44 Turn-On Time (Shutdown) Parameters (t_{EN})

The turn-on time (shutdown) parameter, t_{EN}, is defined as the time from when the turn-on voltage is applied to the shutdown pin to when the supply current has reached half of its final value. It is expressed in units of s.

You need to carefully read the data sheet to understand the logic level of the enable/disable function. Some are referenced to the most negative supply, which can force you to design a logic-level shifting circuit.

Relevance: This parameter is only relevant when the op amp has an enable/disable input, and when this disable/enable function is used.

A.45 Fall Time Parameter (t_f)

The fall time parameter, t_f, is defined as the time required for an output voltage step to change from 90% to 10% of its final value. It is expressed in units of s.

Relevance: Primarily relevant for systems that have a large voltage swing on non-sinusoidal signals.

A.46 Total Harmonic Distortion Parameter (THD)

The total harmonic distortion parameter, THD, is defined as the ratio of the RMS voltage of the harmonics of the fundamental signal to the total RMS voltage at the output. THD is expressed in dBc or %. THD does not account for the noise as does the total harmonic distortion plus noise parameter.

Relevance: This and the following parameter are primarily for audio applications and any application where signal integrity is important.

A.47 Total Harmonic Distortion Plus Noise Parameter (THD + N)

The total harmonic distortion plus noise parameter, THD + N, is defined as the ratio of the RMS noise voltage plus the RMS harmonic voltage of the fundamental signal to the fundamental RMS voltage signal at the output. It is expressed in dBc or %.

THD + N compares the frequency content of the output signal to the frequency content of the input. Ideally, if the input signal is a pure sine wave, the output signal is a pure sine wave. Owing to nonlinearity and noise sources within the op amp, the output is never pure.

To simplify further, THD + N is the ratio of all other frequency components to the fundamental:

$$\text{THD} + \text{N} = \left[\frac{(\Sigma\text{Harmonic voltages} + \text{Noise voltages})}{\text{Fundamental}} \right] \times 100\% \qquad (A.5)$$

A.48 Maximum Junction Temperature Parameter (T_J)

The maximum junction temperature parameter, T_J, is defined as the temperature over which the die may be operated. Some of the other parameters may change with temperature, leading to degraded operation at temperature extremes. T_J is expressed in units of °C.

Relevance: This parameter is only relevant when the op amp dissipates a lot of heat, in high-speed devices such as current feedback amplifiers driving a heavy load, or power op amps. Another application where this is important would be high-temperature environments such as downhole and geothermal.

A.49 Rise Time Parameter (t_r)

The rise time parameter, t_r, is defined as the time required for an output voltage step to change from 10% to 90% of its final value. It is expressed in units of s.

Relevance: Primarily relevant for systems that have a large voltage swing on non-sinusoidal signals.

A.50 Settling Time Parameter (t_s)

The settling time parameter, t_s, is defined as the time required for the output voltage to settle within the specified error band of the final value with a step change at the input. It is also known as total response time, t_{tot}. It is expressed in units of s.

Relevance: Settling time is greatly affected by the application, such as a filter circuit where capacitors can store energy. Therefore, it should be measured in-circuit. It is in particular a design issue in data acquisition circuits when signals are changing rapidly. An example is when using an op amp following a multiplexer to buffer the input to an analog-to-digital converter (ADC). Step changes can occur at the input to the op amp when the multiplexer changes channels. The output of the op amp must settle to within a certain tolerance before the ADC samples the signal.

A.51 Storage Temperature Parameter (T_S or T_{stg})

The storage temperature parameter, T_S or T_{stg}, is defined as the temperature over which the op amp may be stored (unpowered) for long periods without damage. It is expressed in units of °C.

Relevance: This parameter is almost irrelevant, because most op amps will be stored in benign environments. It is of importance for space applications, where interfaces will only be under power when a space probe reaches its destination.

A.52 Supply Voltage Condition (V_{CC} or V_{DD})

Important

The supply voltage condition, V_{CC} or V_{DD}, is defined as the bias voltage applied to the op amp power supply pin(s). For single-supply applications it is specified as a positive value and for split-supply applications it is specified as a $\pm$ value, referenced to analog ground. It is expressed in units of V.

V_{CC} or V_{DD} is often defined in the maximum ratings, recommended operating conditions, and as a test condition in parameter tables and graphs because the voltage supplied has an important impact on the way a circuit operates. It is also used as one of the axis variables in some of the characteristic graphs.

Relevance: All op amp applications.

A.53 Input Voltage Range Condition or Parameter (V_I)

Important

The input voltage range parameter, V_I, is defined as the range of input voltages that may be applied to either the IN + or IN − inputs. The input voltage range condition, V_I, is defined as the voltage delivered to a circuit input when testing for V_O on a graph such as "Large signal inverting pulse response vs. time". It is expressed in units of V for either a condition or a parameter.

Relevance: Just as not all op amps are "rail to rail" on their outputs, neither are they rail to rail on their inputs. This parameter is probably more important than the following three input parameters, but failure to take this input into account can lead to clipping that will look like output clipping. You will encounter this most often in low and unity gain stages when the input signal is large enough to hit this limit.

A.54   Common-Mode Input Voltage Condition (V_{ic})

The common-mode input voltage condition, V_{ic}, is defined as the voltage that is common to both input pins. It is expressed in units of V. V_{io} set at $V_{DD}/2$ (for single-supply op amps) is often used as a condition when testing for various parameters including V_{IO}, I_{IO}, I_{IB}, V_{OH}, and V_{OL}.

Relevance: When a two-wire signal is subject to noise and this noise is being received equally on both signal lines. It can be rejected by a differential amplifier with good common-mode rejection.

A.55   Common-Mode Input Voltage Range Parameter (V_{ICR})

The common-mode input voltage range parameter, V_{ICR}, is defined as the range of common-mode input voltage that, if exceeded, may cause the operational amplifier to cease functioning properly. This sometimes is taken as the voltage range over which the input offset voltage remains within a set limit. It is expressed in units of V.

The input common voltage, V_{IC}, is defined as the average voltage at the inverting and non-inverting input pins. If the common-mode voltage gets too high or too low, the inputs will shut down and proper operation ceases. The common-mode input voltage range, V_{ICR}, specifies the range over which normal operation is guaranteed. The trends toward lower and single supply voltages make V_{ICR} of increasing concern.

Relevance: Rail-to-rail input is required when a non-inverting unity gain amplifier is used and the input signal ranges between both power rails. An example of this is the input of an analog-to-digital converter in a low-voltage, single-supply system. High-side sensing circuits require operation at the positive input rail.

A.56   Differential Input Voltage Parameter (V_{ID})

The differential input voltage parameter, V_{ID}, is defined as the voltage at the non-inverting input with respect to the inverting input. It is expressed in units of V.

V_{ID} is usually defined in the absolute maximum ratings table because stress beyond this limit may cause permanent damage to the device.

Relevance: As an absolute maximum value, it is seldom used in normal operation.

A.57 Differential Input Voltage Range Parameter (V_{DIR})

The input common-mode voltage range parameter, V_{DIR}, is defined as the range of differential input voltage that, if exceeded, may cause the operational amplifier to cease functioning properly. It is expressed in units of V.

Some devices have protection built into them, and the current into the input needs to be limited.

Relevance: Normally, differential input mode voltage limit is not a design issue.

A.58 Turn-On Voltage (Shutdown) Parameter ($V_{IH\text{-}SHDN}$ or $V_{(ON)}$)

The turn-on voltage (shutdown) parameter, $V_{IH\text{-}SHDN}$ or $V_{(ON)}$, is defined as the voltage required on the shutdown pin to turn the device on. It is expressed in units of V.

Relevance: This parameter is only relevant when the op amp has an enable/disable input, and when this disable/enable function is used.

A.59 Turn-Off Voltage (Shutdown) Parameters ($V_{IL\text{-}SHDN}$ or $V_{(OFF)}$)

The turn-off voltage (shutdown) parameter, $V_{IL\text{-}SHDN}$ or $V_{(OFF)}$, is defined as the voltage required on the shutdown pin to turn the device off. It is expressed in units of V.

Relevance: This parameter is only relevant when the op amp has an enable/disable input, and when this disable/enable function is used.

A.60 Input Voltage Condition (V_{IN})

The input voltage condition, V_{IN}, is defined as the DC voltage delivered to a circuit input when testing for V_n. It is expressed in units of V.

Relevance: This condition is only included for completeness.

A.61 Input Offset Voltage Parameter (V_{IO} or V_{OS})

Important

The input offset voltage parameter, V_{IO} or V_{OS}, is defined as the DC voltage that must be applied between the input terminals to cancel DC offsets within the op amp. It is expressed in units of V.

All op amps require a small voltage between their inverting and non-inverting inputs to balance mismatches due to unavoidable process variations. The required voltage is known

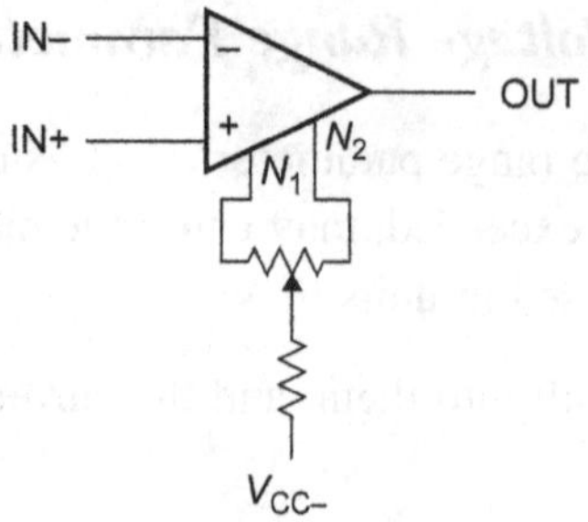

Figure A.5
Offset Voltage Adjust

as the input offset voltage and is abbreviated V_{IO}. V_{IO} is an input referred parameter. This means that it is amplified by the positive closed-loop gain of the circuit.

Input offset voltage is of concern whenever DC accuracy is required of the circuit. One way to null the offset is to use external null inputs on a single op amp package (Figure A.5). A potentiometer is connected between the null inputs with the adjustable terminal connected to the negative supply through a series resistor. The input offset voltage is nulled by shorting the inputs and adjusting the potentiometer until the output is zero.

Relevance: Important for precision DC applications. For AC coupled applications, this parameter may be completely irrelevant.

A.62 Equivalent Input Noise Voltage Parameter (V_n)

Important

The equivalent input noise voltage parameter, V_n, is defined as the internal noise voltage reflected back to an ideal voltage source in parallel with the input pins at a specific frequency. It is expressed in units of $V/\sqrt{Hz}$.

When this parameter is measured, the noise measured at the output (with the input connected to virtual ground) is divided by the gain of the amplifier circuit. This is the amplitude of noise at the input that would be amplified by an ideal amplifier to cause an equivalent signal at the output.

V_n is sometimes defined at several frequencies in the operating characteristics table or as a graph.

Relevance: Important in any application where low noise is important. This probably includes the majority of applications. I do not think I have ever run across an application where the circuit noise can be as high as possible without impacting performance in some way.

A.63 High-Level Output Voltage Condition or Parameter (V_{OH})

Important

The high-level output voltage parameter, V_{OH}, is defined as the positive rail of the op amp output for the load current conditions applied to the power pins. When the V_{OH} parameter is tested it may be defined with I_{OH} of -1 mA, -20 mA, -35 mA, and -50 mA load. When V_{OH} is listed on a data sheet as a test condition it is used for testing another parameter. Whether V_{OH} is a condition or a parameter it is expressed in units of V.

Relevance: I choose to flag this and the companion V_{OL} parameter as important instead of the $V_{OM\pm}$ or the $V_{O(PP)}$ because these are the two that are most commonly used to express the output voltage swing of the op amp. Although they are more important in large signal applications, you should always keep them in mind, especially in DC applications where it is very easy to run out of output range.

A.64 Low-Level Output Voltage Condition or Parameter (V_{OL})

Important

The low-level output voltage parameter, V_{OL}, is defined as the negative rail of the op amp output for the load current conditions applied to the power pins. When the V_{OL} parameter is tested it may be defined with I_{OL} of -1 mA, -20 mA, -35 mA, and -50 mA load. When V_{OL} is listed on a data sheet as a test condition it is used for testing another parameter. Whether V_{OL} is a condition or parameter it is expressed in units of V.

A.65 Maximum Peak-to-Peak Output Voltage Swing Parameter ($V_{OM\pm}$)

The maximum peak-to-peak output voltage swing parameter, $V_{OM\pm}$, is defined as the maximum peak-to-peak output voltage that can be obtained without clipping when the op amp is operated from a bipolar supply. It is expressed in units of V.

Relevance: When this parameter is specified in place of V_{OH} and V_{OL}, it becomes just as important as they are. Because very few data sheets actually specify the output voltage swing this way, I do not flag this as an important parameter.

A.66 Peak-to-Peak Output Voltage Swing Condition or Parameter ($V_{O(PP)}$)

The peak-to-peak output voltage swing condition, $V_{O(PP)}$, is defined as the peak-to-peak voltage set up on the output waveform to test for parameters such as A_{VD} or SR.

The peak-to-peak output voltage swing parameter, $V_{O(PP)}$, is the maximum peak-to-peak output voltage that an op amp can deliver. When it is measured, V_{DD}, THD + H, R_L and T_A are the typical test conditions.

It is expressed in units of V for either a condition or parameter.

Relevance: When this parameter is specified in place of V_{OH} and V_{OL}, it becomes just as important as they are. Because very few data sheets actually specify the output voltage swing this way, I do not flag this as an important parameter.

A.67 Step Voltage Peak-to-Peak Condition ($V_{(STEP)PP}$)

The step voltage peak-to-peak condition, $V_{(STEP)PP}$, is defined as the peak-to-peak voltage step that is used as a test condition for parameters such as t_s. It is expressed in units of V.

Relevance: This is a test condition, so its relevance to a designer is low.

A.68 Crosstalk Parameter (X_T)

The crosstalk parameter, X_T, is defined as the ratio of the change in output voltage of a driven channel to the resulting change in output voltage from another channel that is not driven. It is expressed in units of dB.

X_T is a function of how good the separation is between channels in an IC package or system. It is caused by the signal from one channel being coupled to the other channel inductively, capacitatively, through the power supply, etc.

Relevance: Only relevant when there are two or more op amps per IC package.

A.69 Output Impedance Parameter (Z_o)

The output impedance parameter, Z_o, is defined as the frequency-dependent small-signal impedance that is placed in series with an ideal amplifier and the output terminal in a closed-loop configuration. It is expressed in units of Ω.

Relevance: Often confused with R_o (Section A.38), this includes parasitic inductance and capacitance elements. Of little relevance unless you are driving a heavy load.

A.70 Open-Loop Transimpedance Parameter (Z_t)

The open-loop transimpedance parameter, Z_t, is defined as the frequency-dependent ratio of change in output voltage to the frequency-dependent change in current at the inverting input in a transimpedance or current feedback amplifier. It is expressed in units of Ω.

Relevance: Current feedback and transimpedance amplifier designs.

A.71 Differential Phase Error Parameter (Φ_D)

The differential phase error parameter, Φ_D, is defined as the change in AC phase with change in DC level. The AC signal is 40 IRE (0.28 V PK) and the DC level change is ± 100 IRE (± 0.7 V). It is typically tested at 3.58 MHz (NTSC) or 4.43 MHz (PAL) carrier frequencies. It is expressed in units of degrees, °.

Relevance: With the conversion to digital broadcast video, this parameter is quickly becoming less relevant.

A.72 Phase Margin Parameter (Φ_m)

Important

The phase margin parameter, Φ_m, is defined as the absolute value of the difference in the phase shift of 180° and the phase shift at unity gain. Φ_m is measured open loop and is expressed in units of degrees, °.

$$\Phi_m = 180° - \Phi@B1 \qquad (A.6)$$

Gain margin (A_m) and phase margin (Φ_m) are different ways of specifying the stability of the circuit. Figure A.1 shows Φ_m graphically.

Relevance: This parameter is a measure of the stability of the op amp, and therefore is critical to all op amp circuits. You should always strive for a phase margin no greater than 60°, although circumstances may force you to higher phase margins. By definition, if you reach a phase margin of 180°, the circuit has become an oscillator.

A.73 Bandwidth for 0.1 dB Flatness

Bandwidth for 0.1 dB flatness is defined as the range of frequencies within which the gain is ± 0.1 dB of the nominal value with full output power. It is expressed in units of Hz.

Relevance: This parameter is related to the unity gain bandwidth of the op amp, but in this case it specifies flatness, which is important when the absolute value of the gain is critical at all frequencies.

A.74 Case Temperature for 60 Seconds

The case temperature for 60 seconds is defined as the temperature the case may safely be exposed to for 60 seconds. It is usually specified as an absolute and expressed in °C.

Relevance: This parameter is only relevant when talking about soldering techniques, primarily wave soldering.

A.75 Continuous Total Dissipation Parameter

The continuous total dissipation parameter is defined as the power that can be dissipated by an op amp package, including loads. It is usually specified as an absolute maximum. This parameter may be broken down by ambient temperature and package style in a table. It is expressed in units of W.

Relevance: Do not confuse this with the power supply current. This parameter includes the load. It is probably most relevant in audio power applications, or other applications where continuous high-power operation is required.

A.76 Duration of Short-Circuit Current

The duration of short-circuit current parameter is defined as the amount of time that the output can be shorted to network ground. It is usually specified as an absolute maximum. It is usually expressed in s.

Relevance: This parameter is usually one of the "absolute maximum ratings". It describes a condition that should be considered very undesirable: direct short of the op amp output to ground. With the possible exception of a clumsy individual accidentally shorting the output to ground, I cannot think of a scenario where this parameter would be relevant, or even useful. Keep in mind that the standard pinout of op amps places at least one op amp output adjacent to a power − not ground − pin. This is true for single, dual, and quad op amps. So, accidental short circuits to ground are even less likely if a probe slips; it is much more likely that the op amp output will be shorted to a supply rail, which is usually destructive to the output. This parameter says nothing about shorts to supply rail, only ground. So if you do short an output to a supply, replace the op amp.

A.77 Input Offset Voltage Long-Term Drift Parameter

Input offset voltage long-term drift parameter is defined as the ratio of the change in input offset voltage to the change in time. It is the average value for the month and is expressed in units of V/month.

Relevance: This parameter is important in DC applications, usually transducer interfaces, that will be under continuous power for a long time, and do not have autocalibration techniques. The long-term drift will act as a continually increasing source of error.

A.78 Lead Temperature for 10 or 60 Seconds

The lead temperature for 10 or 60 seconds is defined as the temperature to which the leads may safely be exposed for 10 or 60 seconds. It is usually specified as an absolute maximum and is meant as a guide for automated soldering processes. This parameter is expressed in units of °C.

Relevance: This parameter is only relevant when talking about soldering techniques, primarily wave soldering.

Appendix B: Op Amp Noise Theory

B.1 Introduction

The subject of op amp noise is a complex enough topic that it deserves its own appendix. The purpose of op amp circuitry is the manipulation of the input signal in some fashion. Unfortunately, in the real world, the input signal has unwanted noise superimposed on it.

This discussion of noise is highly theoretical and a bit esoteric. Noise is primarily a function of the semiconductor process used to make an op amp (and the materials used to make passive components). Therefore, there is little, if anything, you can do to eliminate it, other than trying to find different parts until you find one that has acceptable noise limits the way you are going to use it. I have amassed many tidbits of information through the years:

- Many years ago, we discovered by accident that an OP90 had low noise in the ultra-low-frequency range, and used it for that reason.
- The thermal noise characteristics of carbon composition resistors have been well documented. Metal film resistors are better.
- Large-value resistors are noisier than low-value resistors, making low-power/low-noise design difficult.
- Zener diodes are sources of shot noise; if that is a concern, do not use them.
- The EL2125 C is about the lowest noise op amp out there, although newer models have lower noise at higher frequencies.
- You can use the THS4131 high-speed op amp as a low-noise differential amp at audio frequencies.

And so forth. Your job as a designer is to manage noise, and knowing noise theory is one way you can avoid making mistakes and using the wrong component. There are very few things you can do to actually lower the noise of off-the-shelf components, so choose those components carefully. You should prepare a "noise budget" for your design. I use a spreadsheet for the job, which includes the noise of each op amp, the gain of the stage (noise is amplified along with signal), and other things such as the expected signal level compared with the V_{OH} and V_{OL} limits as defined by the power supply and the op amp data sheet. The calculations include a running total of the signal-to-noise ratio, the total noise, and how close I am coming to clipping. I include the data converter in the calculation, because it also defines the clipping level of the signal. I also include power consumption estimates for each stage, so there are few surprises when I actually prototype the circuit. Spreadsheets can be powerful tools for analyzing analog signal chains. I have placed a generic copy of my noise calculation spreadsheet on the companion website for you to use as a starting point for your designs.

B.2 Characterization

Noise is a purely random signal; the instantaneous value and/or phase of the waveform cannot be predicted at any time. Noise can be either generated internally in the op amp, from its associated passive components, or superimposed on the circuit by external sources.

B.2.1 Root Mean Square Versus Peak-to-Peak Noise

Instantaneous noise voltage amplitudes are as likely to be positive as negative. When plotted, they form a random pattern centered on zero. Since noise sources have amplitudes that vary randomly with time, they can only be specified by a probability density function. The most common probability density function is Gaussian. In a Gaussian probability function, there is a mean value of amplitude, which is most likely to occur. The probability that noise amplitude will be higher or lower than the mean falls off in a bell-shaped curve, which is symmetrical around the center (Figure B.1).

σ is the standard deviation of the Gaussian distribution and the root mean square (RMS) value of the noise voltage and current. The instantaneous noise amplitude is within $\pm 1\sigma$ 68% of the time. Theoretically, the instantaneous noise amplitude can have values approaching infinity. However, the probability falls off rapidly as amplitude increases. The instantaneous noise amplitude is within $\pm 3\sigma$ of the mean 99.7% of the time.

σ^2 is the average mean square variation about the average value. This also means that the average mean square variation about the average value, $\overline{i^2}$ or $\overline{e^2}$, is the same as the variance σ^2. Thermal noise and shot noise (see below) have Gaussian probability density functions. The other forms of noise do not.

B.2.2 Noise Floor

When all input sources are turned off and the output is properly terminated, there is a level of noise called the *noise floor* that determines the smallest signal for which the circuit is useful. Your objective is to place the signals that the circuit processes above the noise floor, but below the level where the signals will clip.

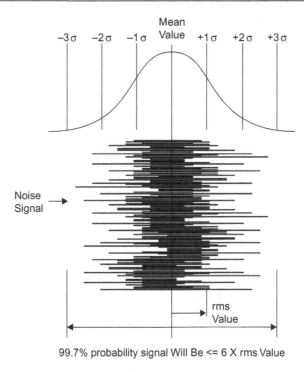

99.7% probability signal Will Be <= 6 X rms Value

Figure B.1

Gaussian Distribution of Noise Energy

B.2.3 Signal-to-Noise Ratio

The noisiness of a signal is defined as its signal-to-noise ratio (SNR):

$$\frac{S_{(f)}}{N_{(f)}} = \frac{\text{RMS Signal Voltage}}{\text{RMS Noise Voltage}} \tag{B.1}$$

In other words, it is a ratio of signal voltage to noise voltage (hence the name *signal-to-noise ratio*). The SNR is commonly used to determine the quality of an audio signal chain: the larger the number, the better. SNR has more applications, though, than just audio. It can apply to any signal to any noise source. The acceptable level of SNR may vary widely from application to application, and may be a system specification.

B.2.4 Multiple Noise Sources

When there are multiple noise sources in a circuit, the total RMS noise signal that results is the square root of the sum of the average mean square values of the individual sources:

$$E_{\text{TotalRMS}} = \sqrt{e_{1\text{RMS}}^2 + e_{2\text{RMS}}^2 + \;\rightleftharpoons e_{n\text{RMS}}^2} \tag{B.2}$$

Put another way, this is the only "break" that the designer gets when dealing with noise. If there are two noise sources of equal amplitude in the circuit, the total noise is not doubled (increased by 6 dB). It only increases by 3 dB. Consider a very simple case, two noise sources with amplitudes of $2\ V_{RMS}$:

$$E_{TotalRMS} = \sqrt{2^2 + 2^2} = \sqrt{8} = 2.83 V_{RMS} \tag{B.3}$$

Therefore, when there are two equal sources of noise in a circuit, the noise is $20 \times \log(2.83/2) = 3.01$ dB higher than if there were only one source of noise, instead of double (6 dB) as would be intuitively expected.

This relationship means that the worst noise source in the system will tend to dominate the total noise. Consider a system in which one noise source is $10\ V_{RMS}$ and another is $1\ V_{RMS}$:

$$E_{TotalRMS} = \sqrt{10^2 + 1^2} = \sqrt{108} = 10.05\ V_{RMS} \tag{B.4}$$

There is hardly any effect from the 1 V noise source at all.

B.2.5 Noise Units

Noise is normally specified as a spectral density in RMS volts or amps per root hertz, $V/\sqrt{Hz}$ or $A/\sqrt{Hz}$. These are not very user-friendly units. A frequency range is needed to relate these units to actual noise levels that will be observed.

For example:

- An op amp with a noise specification of 2.5 $nV/\sqrt{Hz}$ is used over an audio frequency range of 20 Hz to 20 kHz, with a gain of 40 dB. The output voltage is 0 dBV (1 V).
- To begin with, calculate the *root Hz* part: $\sqrt{20{,}000 - 20} = 141.35$.
- Multiplying this by the noise specification: $2.5 \times 141.35 = 353.38$ nV, which is the equivalent input noise (E_{IN}). The output noise equals the input noise multiplied by the gain, which is 100 (40 dB).

The SNR can be now be calculated:

$$\begin{aligned} 353.38\ nV \times 100 &= 35.3\ \mu V \\ SNR\ (dB) = 20 \times log(1\ V \div 35.3\ \mu V) &= 20 \times log(28{,}329) = 89\ dB \end{aligned} \tag{B.5}$$

The op amp is an excellent choice for this application. Remember, though, that passive components and external noise sources can degrade performance. There is also an increase in noise at low frequencies, due to the $1/f$ effect (see below). If you are designing an AC coupled system, you are fortunate: all you have to do is select an op amp where the $1/f$ effect is below the lowest frequency you are interested in.

Table B.1: Noise Colors

Color	Frequency Content
Purple	f^2
Blue	f
White	1
Pink	$1/f$
Red/brown	$1/f^2$

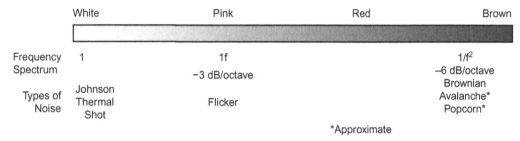

Figure B.2
Noise Colors

B.3 Noise Colors

Op amp noise will appear as the summation of noise types that are related to semiconductor physics, and which are beyond the scope of this book. Noise is broadly categorized by its frequency spectrum, which is related to the frequency by an exponential factor. There is a shorthand way to describe noise, which is called *color*. The colors of noise come from rough analogies to light, and refer to the exponential frequency content as shown in Table B.1.

Just as in the case of light, there is an infinite number of variations between the colors. All exponential powers of frequency are possible. Noise colors above white (exponential increases in content with frequency) are not generally found in op amps, because op amps have internal parasitics and compensation. Therefore, op amp noise appears in the region between white noise and red/brown noise (Figure B.2).

B.3.1 White Noise

White noise is noise in which the frequency and power spectrum is constant and independent of frequency. The signal power for a constant bandwidth (centered at frequency f_o) does not change if f_o is varied. Its name comes from a similarity to white light, which has equal quantities of all colors.

When plotted versus frequency, white noise is a horizontal line of constant value. By definition, white noise would have infinite energy at infinite frequencies. Therefore, white noise always becomes pinkish at high frequencies.

Steady rainfall or radio static on unused channels both approximate a white noise characteristic.

B.3.2 Pink Noise

Pink noise is noise with a $1/f$ frequency and power spectrum excluding DC. It has equal energy per octave (or decade, for that matter). This means that the amplitude decreases logarithmically with frequency. Pink noise is pervasive in nature: many supposedly random events show a $1/f$ characteristic.

B.3.3 Red/Brown Noise

Red noise is named for a connection with red light, which is on the low end of the visible light spectrum. But this noise simulates Brownian motion, so perhaps it should be called brown. Red/brown noise has a -6 dB/octave frequency response and a frequency spectrum of $1/f^2$ excluding DC. Red/brown noise is found in nature. The acoustic characteristics of large bodies of water approximate the red/brown noise frequency response.

B.4 Op Amp Noise

Op amp noise is never specified as white and pink on data sheets. It is specified with a graph of equivalent input noise versus frequency. These graphs usually show two distinct regions:

- lower frequencies where pink noise is the dominant effect
- higher frequencies where white noise is the dominant effect.

Figure B.3 shows a typical data sheet noise plot. It is necessary to break the noise into two parts, the pink part and the white part. In some ways, the noise plot is reminiscent of the op amp open-loop plot, only flipped over and reversed. While this is purely coincidental, it provides you with a new corner frequency, the noise corner frequency f_{nc}. f_{nc} is defined as the point in the frequency spectrum where $1/f$ noise and white noise are equal. Note on the graph in Figure B.3 that the actual noise voltage is higher at f_{nc} owing to the RMS addition of noise sources as defined in Section B.2.4.

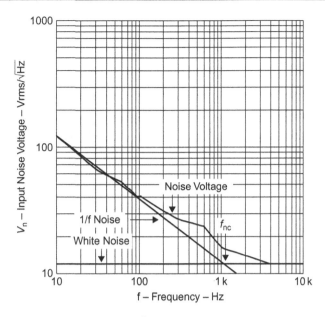

Figure B.3
Typical Op Amp Noise Characteristics

f_{nc} can be determined visually from the graph in Figure B.3. It appears to be 1 kHz. This is done by:

- Taking the white noise portion of the curve, and extrapolating it down to 10 Hz as a horizontal line.
- Taking the portion of the pink noise from 10 Hz to 100 Hz, and extrapolating it as a straight line.
- The point where the two intercept is f_{nc}, the point where the white noise and pink noise are equal in amplitude. The total noise is then $\sqrt{2} \times$ white noise specification (from Section B.2.4). This would be about 17 $nV/\sqrt{Hz}$ for the op amp shown.

B.5 Putting It All Together

I will leave you with a few pointers, some of them repeated from above:

- If you are dealing with an AC coupled system, you should always try to choose an op amp with f_{nc} below the lowest frequency of the bandwidth you are amplifying.
- If you are designing a DC coupled system, you can combat white noise effects by low-pass filtering, or damping the circuit so it will not respond to energetic noise sources higher in frequency than the response time you need.

Table B.2: Noise Calculation Spreadsheet

Noise Calculation Spreadsheet

Input Signal

Min Input Signal	1.0E − 06	V_{PP}
Max Input Signal	1.0E − 03	V_{PP}

Op Amp Stage Parameters

Stage	V_{OL}	V_{OH}	Power Supply Positive V	Power Supply Positive mA	Power Supply Negative V	Power Supply Negative mA	Bandwidth mA	Noise nV/sqrt (Hz)	Frequency Low	Frequency High
1	2	2	15	10	15	10	10	1	970,000	1,030,000
2	2	2	15	10	15	10	10	10	452,500	457,500
3	0.1	0.2	5	2	0	0	0	10	452,500	457,500
4	0	0	5	0	0	0	0	0	452,500	457,500
5	0	0	5	0	0	0	0	0	452,500	457,500
6	0	0	5	0	0	0	0	0	452,500	457,500

Stage	Gain dB	Saturation V_{PP}	Bandwidth Hz	Input Noise nV/sqrt (Hz)	Output min V_{PP}	Output max V_{PP}	Back-off (V_{sat}/V_{OUT}) dB	Output Noise nV	SNR min dB	SNR max dB	Supply Current
1	40	26	60,000	1	1.00E − 04	1.00E − 01	4.83E + 01	2.45E + 04	1.22E + 01	7.22E + 01	20
2	20	26	5000	10	1.00E − 03	1.00E + 00	2.83E + 01	2.55E + 04	3.19E + 01	9.19E + 01	2
3	13	4.7	5000	10	4.47E − 03	4.47E + 00	4.42E − 01	2.57E + 04	4.48E + 01	1.05E + 02	2
4	0	5	5000	0	4.47E − 03	4.47E + 00	9.79E − 01	2.57E + 04	4.48E + 01	1.05E + 02	0
5	0	5	5000	0	4.47E − 03	4.47E + 00	9.79E − 01	2.57E + 04	4.48E + 01	1.05E + 02	0
6	0	5	5000	0	4.47E − 03	4.47E + 00	9.79E − 01	2.57E + 04	4.48E + 01	1.05E + 02	0
ADC	0	5			4.47E − 03	4.47E + 00	9.79E − 01				20

Output Section

Total gain	73	dB
Min output	4.47E − 03	V_{PP}
Max output	4.47E + 00	V_{PP}
SNR min	1.22E + 01	
Supply current	44	mA

- Instabilities mimic blue and purple noise: do not be fooled into thinking you have an op amp with these characteristics. You are dealing with an unstable circuit: make changes to deal with the instabilities!
- Discrete frequencies of noise are coming from outside your circuit. Identify them and where they are coming from, and eliminate them at the source if you can. Implementing a notch filter should be the absolute last resort after trying layout changes, routing changes, and decoupling changes.
- A spreadsheet is a powerful tool for forming an op amp noise budget. One is shown in the design aids section to follow.

B.6 Design Aid: Noise Calculation Spreadsheet

Table B.2 shows the noise calculator spreadsheet that I use. I populated the fields with a hypothetical AM radio example. I assume a 1 μV to 1 mV input signal level. Three stages are used. The first is a 40 dB radiofrequency (RF) gain stage, assuming that a fairly good RF coil and tuning capacitor provide a ±30 kHz bandwidth at 1 MHz. I assume a lossless mixer after this stage and two intermediate-frequency (IF) stages, each one with a bandwidth of 5 kHz. The first IF has a gain of 20 dB and the second a gain of 13 dB, which will produce close to the maximum range at the input of the analog-to-digital converter (ADC).

The first op amp is a high-speed variety with noise of 1 $nV/\sqrt{Hz}$. It consumes 10 mA off each power rail. Assuming that the speed would not reduce, and the part supports it, it could be operated off ±5 V. You should experiment on the actual spreadsheet to see the impact of changes like this on the clipping level (back-off).

The second and third op amps are lower speed, lower power devices. The second is assumed to be a rail-to-rail device that will never exceed the input voltage range of the ADC, which is assumed to be 0–5 V.

Current is calculated magnitude only, because the primary concern is total current. You can assume it is split between two rails equally unless you have a lot of single-supply op amps.

The spreadsheet supports up to six stages. Additional stages can be added by adding rows and copying rows. But if you want to use it "as-is", all you have to do is put in zero gain, zero noise, and zero voltage and current.

In this hypothetical example, there is no automatic gain control (AGC), and sensitivity is 1 μV for 12 dB SNR, but there will only be 45 mV of input to the ADC. A real receiver would need AGC.

Appendix C: Circuit Board Layout Techniques

C.1 General Considerations

Prior discussions have focused on how to design op amp circuitry, how to use integrated circuits (ICs), and the use of associated passive components. There is one additional circuit component that must be considered for the design to be a success: the printed circuit board (PCB) on which the circuit is to be located.

C.1.1 The Printed Circuit Board is a Component of the Op Amp Design

Op amp circuitry is analog circuitry, and is very different from digital circuitry. It must be partitioned in its own section of the board, using special layout techniques.

The effects of the PCB become most apparent in radiofrequency (RF) and high-speed analog circuits, but common mistakes described in this chapter can even affect the performance of audio circuits. The purpose of this chapter is to discuss some of the more common mistakes made by designers and how they degrade performance, and provide simple fixes to avoid the problems.

The PCB layout for analog circuitry must be designed such that the effect of the PCB is transparent to the circuit. Any effect caused by the PCB itself should be minimized, so that the operation of the analog circuitry in production will be the same as the performance of the design and prototype.

Long experience in this profession has revealed an extremely disturbing trend: PCB layout being done by specialists in departments dedicated to the task, these specialists being artists, not electrical engineers, and therefore having neither the experience nor the inclination to take RF and analog performance into account when doing their design. They are used to making all the connections from point to point and calling the job done. PCB layout is thought of as being "beneath" an engineer, or a waste of engineering time.

Considering how critical the PCB is, and how interrelated the performance of the circuitry is with the layout, these attitudes must change! If it is the responsibility of an engineer to see to it an analog circuit works on a PCB, and if they are held accountable should the circuit fail, it is a logical next step to say that the engineer should be the one actually doing the PCB layout. Not only is it necessary, but engineers with the skillset should be commended for taking the initiative, not disrespected for "wasting their time", as I have been.

PCB layout software companies are not doing themselves a favor by pricing their products out the reach of engineers. Fortunately, there are free products like Design Spark PCB that have the potential to put engineers back in control of their circuits, at least at the prototype stage. Once the proper layout has been established, the engineer has a layout to pass to the layout department and can say "do it like this and it will work".

Such is my soapbox on the subject of engineers doing PCB layout.

C.1.2 Prototype, Prototype, PROTOTYPE!

Normal design cycles, particularly of large digital boards, dictate the layout of the PCB as soon as possible. The digital circuitry has been simulated, but in most cases, the production PCB itself is the prototype, and may even be sold to a customer. Digital designers can correct small mistakes by implementing *cuts and jumpers*, reprogramming gate arrays or flash memories, and go on to the next project. This is not the case with analog circuitry. Some of the common design mistakes discussed in this chapter cannot be corrected by the cut and jumper method.

I have been the unfortunate recipient of a simple analog circuit designed by another engineer who was accustomed to the cut and jumper method of correcting his mistakes. Not only was the op amp hooked up with inverting and non-inverting inputs reversed, but an RC time constant had to be added to prevent a race condition. Repercussions from these mistakes, and associated rework problems, caused literally hundreds of hours to be lost from a tight production schedule. Prototyping this circuit would have taken less than a day.

C.1.3 Noise Sources

Noise is the primary limitation on analog circuitry performance. Internal op amp noise is covered in Appendix B. Other types of noise include:

* Conducted emissions: noise that the analog circuitry generates through its connections to other circuits. This is usually negligible in analog circuitry, unless it is high power (such as an audio amplifier that draws heavy currents from its power supply).

- Radiated emissions: noise that the analog circuitry generates, or transmits, through the air. This is also usually negligible in analog circuitry, unless it is high frequency such as video.
- Conducted susceptibility: noise from external circuitry that is conducted into the analog circuit through its connections to other circuits. Analog circuitry must be connected to the "outside world" by at least a ground connection, a power connection, an input, and an output. Noise can be conducted into the circuit through all of these paths, as well as any others that are present.
- Radiated susceptibility: noise that is received through the air (or transmitted into the analog circuitry) from external sources. Analog circuitry, in many cases, resides on a PCB that may have high-speed digital logic including digital signal processor (DSP) chips. High-speed clocks and switching digital signals create considerable radiofrequency interference (RFI). Other sources of radiated noise are endless: the switching power supply in a digital system, cellular telephones, broadcast radio and TV, fluorescent lighting, nearby PCs, lightning in thunderstorms, and so on. Even if the analog circuitry is primarily audio in frequency, RFI may produce noticeable noise in the output. Ever hear GSM tones from a cell phone on a piece of audio gear? That is an excellent example of radiated susceptibility.

Any single or combination of the above sources of noise can render a PCB unusable.

C.2 Printed Circuit Board Mechanical Construction

It is important to choose a PCB with the right mechanical characteristics for the application.

C.2.1 Materials: Choosing the Right One for the Application

PCB materials are available in various grades, as defined by the National Electrical Manufacturers Association (NEMA). It would be very convenient for designers if this organization were closely allied with the electronics industry, controlling parameters such as resistivity and dielectric constant of the material. Unfortunately, that is not the case. NEMA is an electrical safety organization, and the different PCB grades primarily describe the flammability, high-temperature stability, and moisture absorption of the board. Therefore, specifying a given NEMA grade does not guarantee electrical parameters of the material. If this becomes critical for an application, consult the manufacturer of the raw board stock.

Laminated materials are designated with flame-resistant (FR) grades. FR-1 is the least flame resistant and FR-5 the most, as described in Table C.1.

Table C.1: Printed Circuit Board Materials

Grade Designation	Material/Comments
FR-1	Paper/phenolic: room temperature punchable, poor moisture resistance
FR-2	Paper/phenolic: suitable for single-sided PCB consumer equipment, good moisture resistance
FR-3	Paper/epoxy: designed for balance of good mechanical and electrical characteristics
FR-4	Glass cloth/epoxy: excellent mechanical and electrical properties
FR-5	Glass cloth/epoxy: high strength at elevated temperatures, self-extinguishing

Do not use FR-1. There are many examples of boards with burned spots, where high-wattage components have heated a section of the board for a period of time. This grade of PCB material has more in common with cardboard than anything else.

FR-4 is commonly used in industrial-quality equipment, while FR-2 is used in high-volume consumer applications. These two board materials appear to be industry standards. Deviating from these standards can limit the number of raw board material suppliers and PCB houses that can fabricate the board because their tooling is already set up for these materials. Nevertheless, there are applications in which one of the other grades may make sense. For very high frequency applications, it may be necessary to consider Teflon or even ceramic board substrate. For high-temperature boards, polyimide is the material of choice. One thing can be counted on, however: the more exotic the board substrate, the more expensive it will be.

In selecting a board material, pay careful attention to the moisture absorption. Just about every desirable performance characteristic of the board will be negatively affected by moisture. This includes the surface resistance of the board, dielectric leakage, high-voltage breakdown and arcing, and mechanical stability. Also, pay attention to the operating temperature. High operating temperatures can occur in unexpected places, such as in proximity to large digital ICs that are switching at high speeds. Be aware that heat rises, so if one of those 500-pin monster ICs is located directly under a sensitive analog circuit, both the PCB and circuit characteristics may vary with the temperature.

C.2.2 Cladding and Plating

Most boards use copper cladding, because it adheres well to the PCB substrate material.

After the board substrate material has been selected, the next decision is how thick to make the copper foil laminated to it. For most applications, 1-ounce copper is sufficient. If the circuit consumes a lot of power, 2-ounce may be better. Avoid ½-ounce copper, because it tends to break between the trace and the pad. Also avoid abrupt changes in traces width, such as a trace intersecting a pad or via (connection between layers of a PCB). "Teardropping" the trace into the pad or via softens the change in width, avoiding a mechanical stress point in thin copper traces. Teardropping is a common function in most layout programs. Another common function in most layout programs is trace arcs instead of orthogonal routing. If you use orthogonal routing, use 135° angles wherever possible, and avoid 90° angles.

C.2.3 How Many Layers are Best?

Depending on the complexity of the overall circuitry being designed, a designer must decide how many layers the PCB should be comprised of.

C.2.3.1 Single-Sided

Very simple consumer electronics are sometimes fabricated on single-sided PCBs, keeping the raw board material inexpensive (FR-1 or FR-2) with thin copper cladding. These designs frequently include many jumper wires, simulating the circuit routing on a double-sided board. This technique is recommended only for low-frequency circuitry. For reasons described below, this type of design is extremely susceptible to radiated noise. It is harder to design a board of this type, because many things can go wrong. Many complex designs have been successfully implemented with this technique, but they require a lot of forethought. Be prepared to get creative if the design demands high-volume, low-cost PCBs.

C.2.3.2 Double-Sided

The next level of complexity is double-sided. Although there are some double-sided FR-2 boards, they are more commonly fabricated with FR-4 material. The increased strength of FR-4 material supports vias better. Doubled-sided boards are easier to route because there are two layers of foil, and it is possible to route signals by crossing traces on different layers. Crossing traces, however, is not recommended for analog circuitry. Wherever possible, the bottom layer should be devoted to a ground plane, and all other signals routed on the top layer. A ground plane provides several benefits:

- Ground is frequently the most common connection in the circuit. Having it continuous on the bottom layer usually makes the most sense for circuit routing.
- It increases the mechanical strength of the board.
- It lowers the impedance of all ground connections in the circuit, which reduces undesirable conducted noise.

- It adds a distributed capacitance to every net in the circuit, helping to suppress radiated noise.
- It acts a shield to radiated noise coming from underneath the board.

C.2.3.3 Multilayer

Double-sided boards, in spite of their benefits, are not the best method of construction, especially for sensitive or high-speed designs. The most common board thickness is 0.062 inches. This separation is too great for full realization of some of the benefits listed above. Distributed capacitance, for example, is very low owing to the separation.

Critical designs call for multilayer boards. Some of the reasons are obvious, some less so:

- They provide better routing for power as well as ground connections. If the power is also on a plane, it is available to all points in the circuit simply by adding vias.
- Other layers are available for signal routing, making routing easier.
- There will be distributed capacitance between the power and ground planes, reducing high frequency noise.
- There is better electromagnetic interference (EMI)/RFI rejection. There is due to the *image plane effect*, which has been known since the time of Marconi. When a conductor is placed close to a parallel conductive surface, most of the high-frequency currents will return directly under the conductor, flowing in the opposite direction. This mirror image of the conductor within the plane creates a transmission line. Since currents are equal and opposite in the transmission line, it is relatively immune to radiated noise. It also couples the signal very efficiently. The image plane effect works equally well with ground and power planes, but they must be continuous. Any gap or discontinuity causes the beneficial effects to quickly vanish. There is more on this in the following paragraphs.
- The overall project cost is reduced for small production runs. Although multilayer boards are more expensive to manufacture, EMI/RFI requirements from the Federal Communications Commission (FCC) or other agencies may require expensive testing of the design. If there are problems, it can force a complete redesign of the PCB, leading to additional rounds of testing. A multilayer PCB can have as much as 20 dB better EMI/RFI performance over a double-layer PCB. If production volumes are going to be small, it makes sense to make a better PCB to begin with than to try to cut costs and take the risk of failing $25,000–$50,000 tests.

C.3 Grounding

Good grounding is a system-level design consideration. It should be planned into the product from the first conceptual design reviews.

C.3.1 The Most Important Rule: Keep Grounds and Powers Separate

Separate grounding for analog and digital portions of circuitry is one of the simplest and most effective methods of noise suppression. One or more layers on multilayer PCBs are usually devoted to ground planes. If the designer is not careful, the analog circuitry will be connected directly to these ground planes. The analog circuitry return, after all, is the same net in the netlist as digital return. Autorouters respond accordingly and connect all of the grounds together, creating a disaster. *After the fact* separation of grounds on a mixed digital and analog board is almost impossible.

Just as important as separating ground and power planes is to separate analog and digital power. This can be done from one power rail a number of ways, the most common being small series resistors from the common power rail to analog and digital rails, then allowing local bypass capacitors to act as a low-pass filter with the series resistance to filter noise.

C.3.2 Other Ground Rules

*   Do not overlap digital and analog ground or power planes (Figure C.1). Place analog power coincident with analog ground, and digital power coincident with digital ground. If any portion of analog and digital planes overlaps, the distributed capacitance between the overlapping portions will couple high-speed digital noise into the analog circuitry. This defeats the purpose of isolated planes. I often put multilayer boards through the "light test" when crosstalk is suspected. No matter how many layers, I should be able to see a gap between digital and analog planes on every layer through the translucent board. If I cannot, something is overlapping that should not be. How much easier it is to do

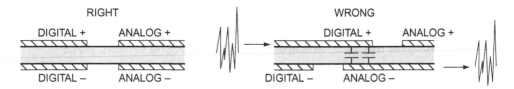

Figure C.1
Digital and Analog Plane Placement

this with the PCB artwork on a viewing program or gerber viewer *before* the board goes out for fabrication!

- Separate grounds does not mean that the grounds are electrically separate on the board. They have to be common at some point, preferably a single, low-impedance point. All ground planes should be connected together at this common connection single point. Often, this point should be located near the connector. But requirements of mixed signal components, such as analog-to-digital converters (ADCs), may force the common connection point to the proximity of the data converters. Other times, the common connection point should be in the power supply circuitry, if separate windings are used on the transformer of a switching supply, for example. There is usually no way to tell in advance which will be better, so pads should be laid out on the board for common connection points in all of those locations. This will allow quick reconfiguration in case of a problem. I have a further design note. In at least one case, ADCs were actually destroyed by ground "bounce" when the analog and digital planes were not connected together right at the converters. In this case, the planes were split right under the component, with the jumper resistor from analog ground to digital ground being right under the converter. That is how critical common grounding points can be!

- It is important to keep digital signals away from analog portions of the circuit. It makes little sense to isolate planes, keep analog traces short, and place passive components carefully if there are high-speed digital traces running right next to the sensitive analog traces. Digital signals must be routed around analog circuitry, and not overlap analog ground and power planes. If not, the design will include a new schematic symbol shown in Figure C.2: the broadcasting antenna!

- Most digital clocks are high enough in frequency that even small capacitances between traces and planes can couple significant noise. Remember that it is not only the fundamental frequency of the clock that can cause a potential problem, but also the higher frequency harmonics.

- It is a good idea to locate analog circuitry as close as possible to the input/output (I/O) connections of the board. Digital designers, used to high-current ICs, will be tempted to make a 50 mil trace run several inches to the analog circuitry thinking that reducing the resistance in the trace will help to get rid of noise. What they have actually done is create a long, skinny capacitor that couples noise from digital ground and power planes into the op amp, making the problem worse! If this is an absolute requirement of the system, place a corresponding "keep out" area on every other layer of the board so that there is no possibility of digital planes or signals crossing the analog power traces and/or planes.

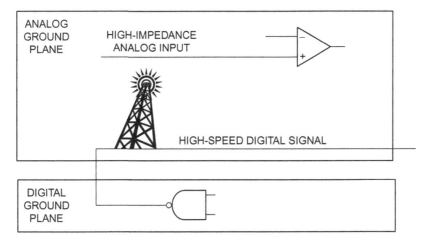

Figure C.2
Broadcasting from PCB Traces

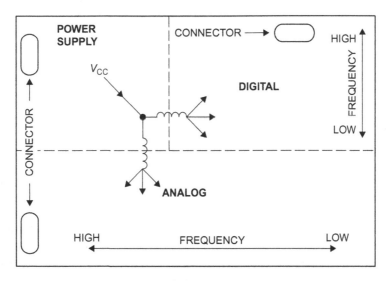

Figure C.3
A Careful Board Layout

C.3.3 A Good Example

Figure C.3 shows one possible board layout. In this system, all electronics, including the power supply, reside on one PCB. Three separate and isolated ground/power planes are employed: one for power, one for digital, and one for analog. Power and ground connections from digital and analog sections of the board are combined only in the supply section, and are combined in close proximity.

High-frequency conducted noise on the power lines is limited by inductors (chokes). In this case, the designer has even located low-frequency analog circuitry close to low-speed digital, keeping high-frequency digital and analog physically apart on the board. This is a good, careful design that has a high likelihood of success, provided that good layout and decoupling rules are also followed.

C.4 Frequency Characteristics of Passive Components

Choosing the right passive components for an analog design is important. In most cases, a *right* passive component will fit on the same pads as a *wrong* passive component, but not always. Start the design process by carefully considering the high-frequency characteristics of passive components, and putting the correct part outline on the board from the start.

Be aware of the frequency limitations of any passive components you use in analog circuitry. Passive components have limited frequency ranges, and operation of the part outside that range can have some very unexpected results. One might think that this discussion only applies to high-speed analog circuits; but high frequencies that are radiated or conducted into a low-speed circuit will affect passive components as well. For example, a simple op amp low-pass filter may well turn into a high-pass filter at radio frequencies.

C.4.1 Resistors

High-frequency performance of resistors is approximated by the schematic shown in Figure C.4.

Resistors are typically one of three types: wire-wound, carbon composition, and film. It does not take a lot of imagination to understand how wire-wound resistors can become inductive because they are coils of resistive wire. Most designers are not aware of the internal construction of film resistors, which are also coils of thin metallic film. Therefore, film resistors are also inductive at high frequencies. The

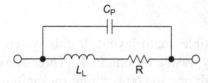

Figure C.4
Resistor High-Frequency Model

inductance of film resistors is lower, however, and values under 2 kΩ are usually suitable for high-frequency work.

The end caps of resistors are parallel, and there will be an associated capacitance. Usually, the resistance will make the parasitic capacitor so "leaky" that the capacitance does not matter. For very high resistances, the capacitance will appear in parallel with the resistance, lowering its impedance at high frequencies.

C.4.2 Capacitors

High-frequency performance of capacitors is approximated by the schematic shown in Figure C.5.

Capacitors are used in analog circuitry for power supply decoupling, as filter components, and as stage coupling components. For an ideal capacitor, reactance decreases by the formula:

$$X_C = \frac{1}{(2\pi f C)} \tag{C.1}$$

where X_C = capacitive reactance (Ω), F = frequency (Hz), and C = capacitance (microfarads, μF).

Therefore, a 10 μF electrolytic capacitor has a reactance of 1.6Ω at 10 kHz, and 160 μΩ at 100 MHz. Right?

In reality, one will never see the 160 μΩ with the electrolytic capacitor. Film and electrolytic capacitors have layers of material wound around each other, which creates a parasitic inductance. Self-inductance effects of ceramic capacitors are much smaller, giving them a higher operating frequency. There is also some leakage current from plate to plate, which appears as a resistance in parallel with the capacitor, as well as resistance within the plates themselves, which add a parasitic series resistance.

Capacitors used for critical analog and RF circuitry should preferentially be the highly stable and low temperature coefficient COG/NPO type or even silver mica.

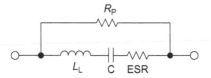

Figure C.5
Capacitor High-Frequency Model

These are vastly superior for systems that will be subjected to wide variation in temperature or to temperature extremes. Unfortunately, these capacitors are limited to lower capacitance values, with capacitors in these dielectrics becoming large, expensive, and hard to obtain in larger values.

The next level in quality is usually X7R dielectric. These can vary widely in capacitance over temperature, particularly above 125°C. Larger values of ceramic capacitors available in X7R have become available, owing to demands from switching power supply designers. X7R capacitors have much lower values of equivalent series resistance (ESR) than electrolytic varieties, meaning that a much lower value can be used for equivalent ripple reduction. But the most common failure mode under conditions of vibration and shock is breakage, which can lead to shorts, so exercise caution. The ESR of ceramic capacitors can be so low that it causes some voltage regulator circuits to become unstable, so small series resistors are sometimes used to add ESR back in.

C.4.3 Inductors

High-frequency performance of inductors is approximated by the schematic shown in Figure C.6.

Inductive reactance is described by the formula:

$$X_{\mathrm{L}} = 2\pi fL \tag{C.2}$$

where X_{L} = inductive reactance (Ω), F = frequency (Hz), and L = inductance (henrys, H).

Therefore, a 10 mH inductor has a reactance of 628Ω at 10 kHz, which increases to 6.28 MΩ at 100 MHz. Right?

In reality, one will never see the 6.28 MΩ with this inductor. Parasitic resistances are easy to understand: the inductor is constructed of wire, which has a given resistance per unit length. Parasitic capacitance is harder to visualize, unless one considers the fact that each turn of wire in the inductor is located next to adjacent

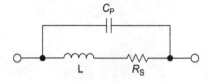

Figure C.6
Inductor High-Frequency Model

turns, forming a capacitor. This parasitic capacitance limits the upper frequency of this inductor to under 1 MHz. Even small wire-wound inductors start to become ineffective in the 10–100 MHz range.

C.4.4 Unexpected Printed Circuit Board Passive Components

In addition to the obvious passive components above, the PCB itself has characteristics that form components every bit as real as those discussed previously, just not as obvious.

C.4.4.1 Printed Circuit Board Trace Characteristics

The layout pattern on a PCB can make it susceptible to radiated noise. A good layout is one that minimizes the susceptibility of analog circuitry to as many radiated noise sources as possible. Unfortunately, there is always a level of RF energy that will be able to upset the normal operation of the circuit. If good design techniques are followed, that level will be one that the circuit never encounters in normal operation.

C.4.4.2 Trace Antennas

A board is susceptible because the pattern of traces and component leads forms antennas. Antenna theory is a complex subject, well beyond the scope of this book. Nevertheless, a few basics are presented here.

Whip Antennas

One basic type of antenna is the whip, or straight conductor. This antenna works because a straight conductor has parasitic inductance, and therefore can concentrate flux from external sources. The impedance of any straight conductor has a resistive and an inductive component:

$$Z = R + j\omega L \tag{C.3}$$

For DC and low frequencies, resistance is the major factor. As the frequency increases, however, the inductance becomes more important. Somewhere in the range of 1–10 kHz, the inductive reactance exceeds the resistance, so the conductor is no longer a low-resistance connection, but, rather, an inductor.

The formula for the inductance of a PCB trace is:

$$L(\mu H) = 0.0002X \cdot \left[\ln\left(\frac{2X}{W+H}\right) + 0.2235\left(\frac{W+H}{X}\right) + 0.5 \right] \tag{C.4}$$

where X = length of the trace, W = width of the trace, and H = thickness of the trace.

The inductance is relatively unaffected by the diameter, since it varies as the logarithm of the circumference. Common wires and PCB traces vary between 6 and 12 nH/cm.

For example, a 10 cm PCB trace has a resistance of 57 mΩ and an inductance of 8 nH/cm. At 100 kHz, the inductive reactance reaches 50 mΩ. At frequencies above 100 kHz, the trace is inductive, not resistive.

A rule of thumb for whip antennas is that they begin to couple significant energy at about 1/20 of the wavelength of the received signal, peaking at ¼ the wavelength. Therefore, the 10 cm conductor of the previous paragraph will begin to be a fairly good antenna at frequencies above 150 MHz. Remember that although the clock generator on a digital PCB may not be operating at a frequency as high as 150 MHz, it approximates a square wave. Square waves will have harmonics throughout the frequency range where PCB conductors become efficient antennas. If through-hole components are mounted in a way that leaves significant lead length, those component leads also become antennas, particularly if they are bent, which introduces the next topic: loop antennas.

Loop Antennas

Another major type of antenna is the loop. The inductance of a straight conductor is dramatically increased by bending it into partial or complete loops. Increased inductance lowers the frequency at which the conductor couples radiated signals into the circuit.

Without realizing it, most digital designers are well versed in loop antenna theory. They know not to make loops in critical signal pathways. Some designers, however, who would never think of making a loop with a high-speed clock or reset signal, will turn right around and create a loop by the technique they use for layout of the analog section of the board. Loop antennas constructed as loops of PCB traces are easy to visualize. What is not as obvious is that slot antennas are just as efficient. Consider the three cases shown in Figure C.7.

Version A is a poor design. It does not use an analog ground plane at all. A loop is formed by the ground and signal traces. An electric field E and perpendicular magnetic field H are created, and form the basis of a loop antenna. A rule of thumb for loop antennas is that the length of each leg is equal to half the most efficiently received wavelength. Remember, however, that even at 1/20 of the wavelength, the loop will still be a fairly efficient antenna.

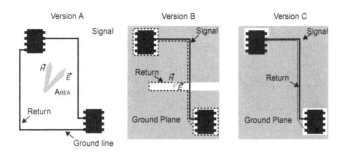

Figure C.7
Loop and Slot Antenna Board Trace Layouts

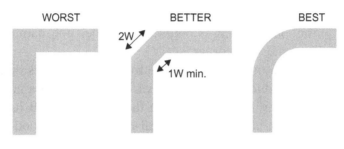

Figure C.8
PCB Trace Corners

Version B is a better design, but there is intrusion into the ground plane, presumably to make room for a signal trace.

Version C is the best design. Signal and return are most coincident with each other, eliminating loop antenna effects completely.

C.4.4.3 Trace Reflections

Reflections and matching are closely related to loop antenna theory, but different enough to warrant their own discussion.

It is a given that not all PCB traces can be straight, and so they will have to turn corners. Figure C.8 shows progressively better techniques of rounding corners. Most computer-aided design (CAD) systems have all of these routing methods available.

- Sharp 90° corners are used to facilitate high-density digital routing. When a PCB trace turns a corner at a 90° angle, a reflection can occur. This is primarily due to the change of width of the trace. At the apex of the turn, the trace width is increased to 1.414 times its width. This upsets the transmission line

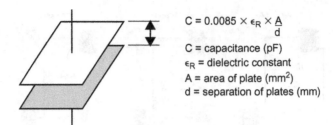

$$C = 0.0085 \times \epsilon_R \times \frac{A}{d}$$

C = capacitance (pF)
ϵ_R = dielectric constant
A = area of plate (mm^2)
d = separation of plates (mm)

Figure C.9
PCB Trace-to-Plane Capacitance Formula

characteristics, especially the distributed capacitance and self-inductance of the trace, resulting in the reflection.

- Although 45° orthogonal routing is better, it still does not maintain constant width as the trace rounds the corner.
- Rounded corners maintain the width of the trace as it changes direction, minimizing reflections due to trace width variation. This does not stop loop antenna effects, however. A suggestion for the advanced PCB layout engineer is to leave rounding to the last step before tear-dropping and flood-filling. Otherwise, the CAD program will slow down doing numerical calculations as the traces are moved around during routing.

C.4.4.4 Trace-to-Plane Capacitors

Since PCB traces are composed of foil, they form capacitance with other traces that they cross on other layers. For two traces crossing each other on adjacent planes, this is seldom a problem. Coincident traces (those that occupy the same routing on different layers) form a long, skinny capacitor. The formula for capacitance is shown in Figure C.9.

C.4.4.5 Trace-to-Trace Capacitors and Inductors

PCB traces are not infinitely thin. They have some finite thickness, as defined by the *ounce* parameter of the copper-clad foil. The higher the number of ounces, the thicker the copper. If two traces run side by side, then there will be capacitative and inductive coupling between them (Figure C.10). The formulas for these parasitic effects can be found in transmission line and/or microstrip references, but are too complex for inclusion here.

Signal lines should not be routed parallel to each other, unless transmission line or microstrip effects are desired. Otherwise, a gap of at least three times the signal trace width should be maintained.

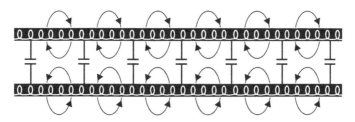

Figure C.10
Coupling Between Parallel Signal Traces

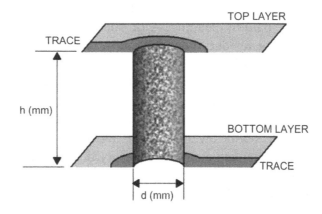

Figure C.11
Via Inductance Measurements

Capacitance between traces in an analog design can become a problem if fixed resistors in the design are large (several megaohms). Capacitance between the inverting and non-inverting inputs of an op amp could easily cause oscillation.

C.4.4.6 Inductive Vias

Whenever routing constraints force a via (Figure C.11), a parasitic inductor is also formed. At a given diameter (*d*) the approximate inductance (*L*) of a via at a height of *h* may be calculated as follows:

$$L = \ \approx \frac{h}{5} \times \left(1 + \ln\left(\frac{4h}{d}\right) \right) \text{nH} \tag{C.5}$$

One of the best methods of combating via capacitance is to route as many signals as possible without vias. This leads to one of my PCB layout truisms: time spent on placement is never wasted. I actually spend a lot of my placement time on the schematic: arranging subcircuits neatly and logically together, using schematic symbols that approximate the true appearance of the part, and do not try to functionally group pins together on large ICs. If the schematic can be drawn that minimizes crossovers, vias and routing on other layers will also be minimized on the PCB layout, because you have already developed strategies to avoid it. I can truthfully say that a lot of my placement and routing time is actually done on the schematic, making placement and routing a cinch on the PCB.

C.4.4.7 Flux Residue Resistance

Yes, even an unclean board can affect analog circuit performance. Be aware that if the circuit has very high resistances — even in the low megaohms — special attention may need to be paid to cleaning. A finished assembly may be adversely affected by flux or cleaning residue. The electronics industry in the past few years has joined the rest of the world in becoming more environmentally responsible. Hazardous chemicals are being removed from the manufacturing process, including flux that has to be cleaned with organic solvents. Water-soluble fluxes are becoming more common, but water itself can become contaminated easily with impurities. These impurities will lower the insulation characteristics of the PCB substrate. It is vitally important to clean with freshly distilled water every time a high-impedance circuit is cleaned. There are applications that may call for the older organic fluxes and solvents, such as very-low-power battery-powered equipment with resistors in the tens of megaohms. Nothing can beat a good vapor defluxing machine for ensuring that the board is clean.

C.5 Decoupling

Noise can propagate into analog circuitry through the power pins of the circuit as a whole and op amp itself. Bypass capacitors are used to reduce the coupled noise by providing low-impedance power sources local to the analog circuitry.

C.5.1 Digital Circuitry: A Major Problem for Analog Circuitry

If analog circuitry is located on the same board as digital circuitry, it is important to understand a little about the electrical characteristics of digital gates. A typical digital output consists of two transistors connected in series between power and ground (Figure C.12). One transistor is turned on and the other turned off to produce logic high, and vice versa for logic low. Because one transistor is turned off for either logic state, the power consumption for either logic state is low, while the gate is static at that level.

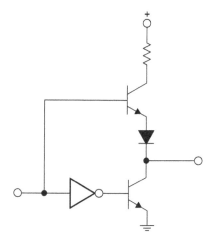

Figure C.12
Typical Logic Gate Output Structure

The situation changes dramatically whenever the output switches from one logic state to the other. There may be a brief period when both transistors conduct simultaneously. During this period, current drawn from the power supply increases dramatically, since there is now a low-impedance path through the two transistors from power to ground. Power consumption rises dramatically and then falls, creating a droop on the power supply voltage, and a corresponding current spike. The current spike will radiate RF energy. There may be dozens, hundreds, or even thousands of such outputs on a PCB, so the aggregate effect may be quite dramatic. It is impossible to predict the frequencies of these spikes. Digital switching noise will be broadband, with harmonics throughout the spectrum. A general rejection technique is required, rather than one that rejects a specific frequency.

C.5.2 Choosing the Right Capacitor

Table C.2 is a rough guideline describing the maximum useful frequencies of common capacitor types.

Obviously from Table C.2, tantalum electrolytic capacitors are useless for frequencies above 1 MHz. Effective high-frequency decoupling at higher frequencies demands a ceramic capacitor. Self-resonances of the capacitor must be known and avoided or the capacitor may not help, or may even make the problem worse. Figure C.13 illustrates the typical self-resonance of two capacitors commonly used for bypassing: 10 μF tantalum electrolytic and 0.01 μF ceramic.

Table C.2: Recommended Maximum Frequencies for Capacitors

Type	Max. Frequency
Aluminum electrolytic	100 kHz
Tantalum electrolytic	1 MHz
Mica	500 MHz
Ceramic	1 GHz

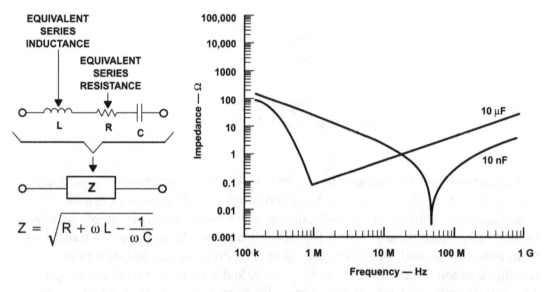

Figure C.13
Capacitor Self-Resonance

Consider these resonances to be typical values; the characteristics of actual capacitors can vary from manufacturer to manufacturer and from grade of part to grade of part. The important thing is to make sure that the self-resonance of the smallest capacitor occurs at a frequency above the range of the noise that must be rejected. Otherwise, the capacitor will enter a region where it is inductive.

C.5.3 Decoupling at the Integrated Circuit Level

The method most often used to decouple the high-frequency noise is to include a capacitor, or multiple capacitors connected from the op amp power pin to the op amp ground pin. It is important to keep the traces on this decoupling capacitor short. If not, the traces on the PCB will have significant self-inductance, defeating the purpose of the capacitor.

A decoupling capacitor must be included on every op amp package, whether it contains one, two, or four devices per package. The value of the capacitor must be picked carefully to reject the type of noise present in the circuit.

In particularly troublesome cases, it may be necessary to add a series resistor in a 10–100Ω range into the power supply line connecting to the op amp. This resistor is in addition to the decoupling capacitors, which are the first line of defense. The resistor should be located before, not after, the capacitors. The resistor forms a low-pass filter with the decoupling capacitors. There is a penalty to pay for this technique: depending on the power consumption of the op amp, it will reduce the rail-to-rail voltage range. The resistor forms a voltage divider with the op amp as a resistive active component in the lower leg of the divider. Depending on the application, this may or may not be acceptable.

C.5.4 Decoupling at the Board Level

There is usually enough low-frequency ripple on the power supply at the board input to warrant a bulk decoupling capacitor at the power input. This capacitor is used primarily to reject low-frequency signals, so an aluminum or a tantalum capacitor is acceptable. An additional ceramic cap at the power input will decouple any stray high-frequency switching noise that may be coupled off the other boards.

C.6 Input and Output Isolation

Many noise problems are the result of noise being conducted into the circuit through its input and output pins. Owing to the high-frequency limitations of passive components, the response of the circuit to high-frequency noise may be quite unpredictable.

In situations in which conducted noise is substantially different in frequency from the normal operating range of the circuit, the solution may be as simple as a passive RC low-pass filter that rejects RF frequencies while having negligible effect at audio frequencies. A good example is RF noise being conducted into an audio op amp circuit.

The effect of radiated energy coupling into an analog circuit can be so bad that the only solution to the problem may be to completely shield the circuit from radiated energy. This shield is called a *Faraday cage*, and must be carefully designed so that frequencies that are causing the problem are not allowed to enter the circuit. This means that the shield must have no holes or slots larger than 1/20 the wavelength of the offending frequency.

Figure C.14
PCB Shield

Figure C.14 shows a good example of a PCB shield. It has to have holes to allow access to adjustment points, but those holes are too small to allow interference at the frequencies of interest (AM and FM radio). It is a good idea to design a PCB from the beginning to have enough room to add a metal shield if it becomes necessary. If a shield is used, frequently the problem will be severe enough that ferrite beads will also be required on all connections to the circuit.

C.7 Packages

Op amps are commonly supplied one, two, or four per package. Single op amps often contain additional inputs for features such as offset nulling. Op amps supplied two and four per package offer only inverting and non-inverting inputs, and the output. If the additional features are important, the only package choice is single. Be aware, though, that the offset-nulling pins on a single op amp package can act as secondary inputs, and must be treated carefully. Consult the data sheet on the particular device being used. Common op amp pinouts are shown in Figure C.15.

The single op amp package places the output on the opposite side from the inputs. This can be a disadvantage at high speeds, because it forces longer PCB traces. Some high-speed amplifiers are now adding a second connection to the output on pin 1, so feedback path length can be shortened.

It is popular to use dual op amps for stereophonic circuits, and quad op amps for filter stages with many sections. There is a penalty for doing so, however. Although modern processing techniques provide high levels of isolation between amplifiers

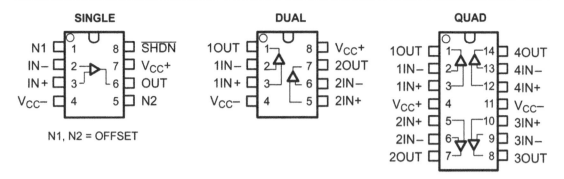

Figure C.15
Common Op Amp Pinouts

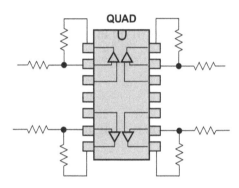

Figure C.16
Mirror-Image Layout for Quad Op Amp Package

on the same piece of silicon, there will be some crosstalk. If isolation between amplifiers is important, then single packages should be considered. Crosstalk problems are not limited to the IC; the dual and quad packages place a high density of passive components in close proximity to each other. This proximity will lead to some crosstalk.

Dual and quad op amp packages offer some additional benefits beyond density. The amplifier stages tend to be mirror images of each other. If similar stages are to be laid out on the PCB, the layout only needs to be done once; then it can be mirror-imaged to form the other stage. Figure C.16 illustrates this effect for four inverting op amp stages implemented in a quad package.

These illustrations, however, do not show all connections required for operation, in particular, the half-supply generator for single-supply operation. Modifying the

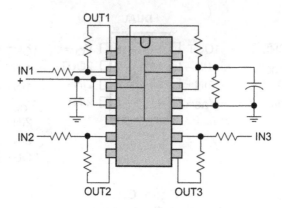

Figure C.17
Quad Op Amp Package Layout with Half-Supply Generator

diagram of Figure C.16 to use the fourth op amp as a half supply generator is shown in Figure C.17.

C.8 Summary

Keep the following points in mind when designing a PCB for analog circuitry.

C.8.1 General

- Think of the PCB as a component of the design.
- Know and understand the types of noise the circuit will be subjected to.
- Prototype the circuit.

C.8.2 Board Structure

- Use a high-quality board material such as FR-4.
- Multilayer boards are as much as 20 dB better than double-sided boards.
- Use separate, non-overlapping ground and power planes.
- Place power and ground planes to the interior of the board instead of the exterior.

C.8.3 Components

- Be aware of frequency limitations of traces and other passive components.
- Use surface mount for high-speed analog circuitry.
- Keep traces as short as possible.

- Use narrow traces if long traces are required.
- Terminate unused op amp sections properly.

C.8.4 Routing

- Never route digital traces through analog sections of the board, or vice versa.
- Make sure that traces to the inverting input of the op amp are short.
- Make sure that traces to the inverting and non-inverting inputs of the op amp do not parallel each other for any significant length.
- It is better to avoid vias, but the self-inductance of vias is small enough that a few should cause few problems.
- Do not use right-angle traces; use curves if at all possible.

C.8.5 Bypass

- Use the correct type of capacitor to reject the conducted frequency range.
- Use tantalum capacitors at power input connectors for filtering power supply ripple.
- Use ceramic capacitors at power input connectors for high-frequency conducted noise.
- Use ceramic capacitors at the power connections of every op amp IC package. More than one capacitor may be necessary to cover different frequency ranges.
- Change the capacitor to a smaller value, not a larger one, if oscillation occurs.
- Add series resistors for stubborn cases.
- Bypass analog power only to analog return, never to digital return.

Appendix D: Op Amp Circuit Collection

D.1 Introduction

This appendix has just a few op amp circuits that do not fit well into other sections of this book, in the grand tradition of "Floobydust" — a section of an old, out-of-print National Semiconductor publication. It is hoped that these circuits can give you a starting point for other interesting op amp applications, and give a glimpse into just what an op amp can really do.

D.2 Simulated Inductor

The circuit in Figure D.1 reverses the operation of a capacitor, thus making a simulated inductor. An inductor resists any change in its current, so when a DC voltage is applied to an inductance, the current rises *slowly*, and the voltage falls as the external resistance becomes more significant.

$$L = R_1 \times R_2 \times C_1 \tag{D.1}$$

An inductor passes low frequencies more readily than high frequencies, the opposite of a capacitor. An ideal inductor has zero resistance. It passes DC without limitation, but it has infinite impedance at infinite frequency.

If a DC voltage is suddenly applied to the inverting input through resistor R_1, the op amp ignores the sudden load because the change is also coupled directly to the non-inverting input via C_1. The op amp represents high impedance, just as an inductor does.

As C_1 charges through R_2, the voltage across R_2 falls, so the op amp draws current from the input through R_1. This continues as the capacitor charges, and eventually the op amp has an input and output close to virtual ground ($V_{CC}/2$).

When C_1 is fully charged, resistor R_1 limits the current flow, and this appears as a series resistance within the simulated inductor. This series resistance limits the Q of the inductor. Real inductors generally have much less resistance than the simulated variety.

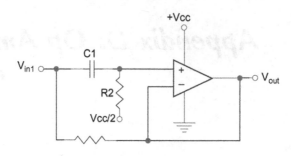

Figure D.1
Simulated Inductor

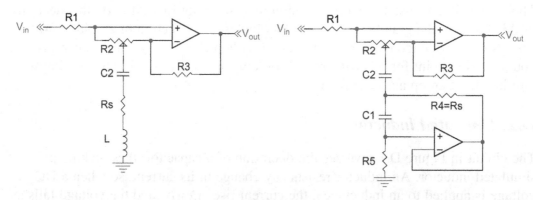

Figure D.2
Graphic Equalizer

There are some limitations to a simulated inductor:

- One end of the inductor is connected to virtual ground.
- The simulated inductor cannot be made with high Q, owing to the series resistor R_1.
- It does not have the same energy storage as a real inductor. The collapse of the magnetic field in a real inductor causes large voltage spikes of opposite polarity. The simulated inductor is limited to the voltage swing of the op amp, so the flyback pulse is limited to the voltage swing.

These limitations limit the use of simulated inductors, but there is one application that is perfect for simulated inductors: graphic equalizers.

To make a graphic equalizer, start with the basic op amp circuit shown in Figure D.2. The inductor L is shown with a parasitic resistance R_S. It resonates with C_2, and depending on the setting of potentiometer R_2, the stage produces either a gain or a loss at the resonant frequency. The parasitic resistance of the inductor R_S also sets the Q of the resonant circuit. Therefore, it will determine the number of stages of equalization required to cover the audio band. On the right-hand side of Figure D.2, the inductor L has been replaced by a simulated inductor circuit. To form the graphic equalizer, multiple stages of equalization are added in parallel by placing more potentiometers in parallel with R_2.

It may be difficult for a home hobbyist to construct a graphic equalizer based on these circuits, because graphic equalizers require an unusual double taper slide potentiometer. Printed circuit board (PCB) parasitics will also make construction of third octave multiband graphic equalizers hard.

D.3 Constant Current Generator

A circuit for a constant current generator is shown in Figure D.3.

Convenient current reference up to 20 mA:

$$I = \frac{V_Z}{R_2} = \frac{6}{300} = 20 \ \text{mA}$$
$$R_1 = \frac{15 - V_Z}{I_Z} = \frac{9}{25} = 360\Omega \qquad \text{(D.2)}$$
$$R_{L \ min} = \frac{\text{Saturation Voltage}}{I} = \frac{13.5 \ \text{V}}{20 \ \text{mA}} = 675\Omega$$

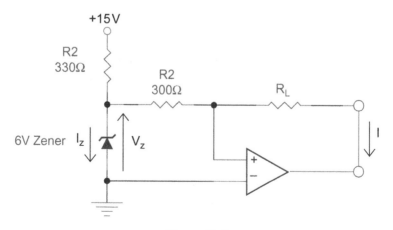

Figure D.3
Constant Current Generator

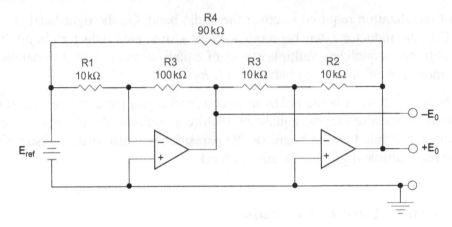

Figure D.4
Reference Voltage Supply

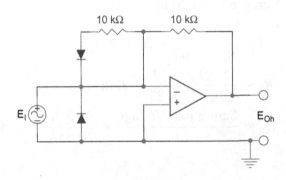

Figure D.5
Absolute Value Circuit

D.4 Inverted Voltage Reference

The circuit of Figure D.4 can be used to generate a negative voltage reference equal to a positive reference voltage. This circuit requires split supplies, however.

D.5 Absolute Value

An absolute value circuit is shown in Figure D.5.

$$
\begin{aligned}
&+E_I \ \ \text{follower circuit} \\
&-E_I \ \ \text{inverter circuit} \\
&E_O = |E_I|
\end{aligned}
\tag{D.3}
$$

Full wave rectification is achievable. Reverse the diodes to give $E_O = -|E_I|$.

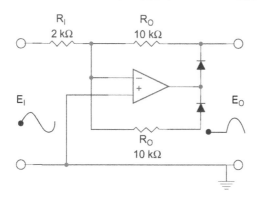

Figure D.6
Precision Rectifier

D.6 Precision Rectifier

A precision rectifier circuit is shown in Figure D.6.

$$E_{O \text{ peak}} = \frac{-R_O}{R_I} E_{I \text{ peak}} = -5E_{I \text{ peak}} \tag{D.4}$$

Half wave with amplification is possible if desired. Placing rectifiers in a feedback loop decreases non-linearity to a very small value.

D.7 AC to DC Converter

An AC to DC converter circuit is shown in Figure D.7.

$$E_{O \text{ average}} = 0.9E_I\text{rms}$$
$$E_I = 6 \text{ mV to 6 V rms @ 10 to 1000 Hz} \tag{D.5}$$

This allows precision conversion for measurement or control. It is a full wave rectifier with a smoothing filter.

D.8 Full Wave Rectifier

A full wave rectifier circuit is shown in Figure D.8. This is a precision absolute value circuit.

D.9 Tone Control

One rather unusual op amp circuit is the tone control circuit of Figure D.9. It bears some superficial resemblance to the "Twin-T" circuit configuration, but it is not a

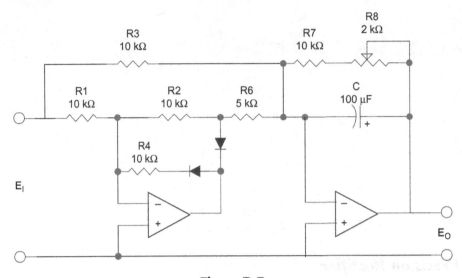

Figure D.7
AC to DC Converter

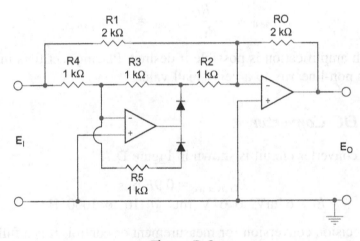

Figure D.8
Full Wave Rectifier

Twin-T topology. It is actually a hybrid of one-pole low-pass and high-pass circuits with gain and attenuation.

The mid-range for the tone adjustments is 1 kHz. It gives about ± 20 dB of boost and cut for bass and treble. The circuit is a minimum component solution, seeking to limit cost. This circuit, unlike other similar circuits, uses linear potentiometers instead of logarithmic ones. Two different potentiometer values are unavoidable, but the capacitors are the same value except for the coupling capacitor. The ideal value of capacitor is 0.016 μF, which is an E-24 value, so the more common E-12

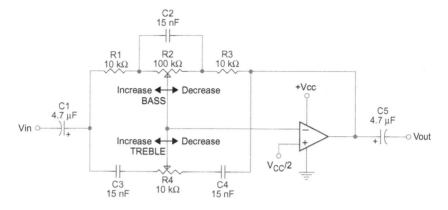

Figure D.9
Tone Control

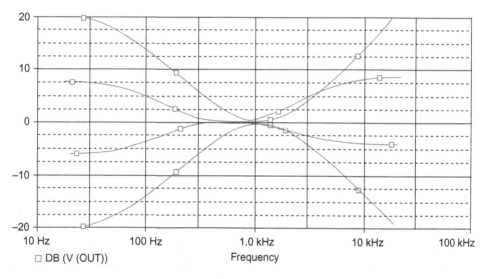

Figure D.10
Tone Control Response

value of 0.015 μF is used instead. Even that value is a bit odd, but it is easier to find an oddball capacitor value than it is an oddball potentiometer value.

Figure D.10 shows the response of the circuit with the potentiometers at the extremes, and at ¼ and ¾ positions. The mid-position, although not shown, is flat to within a few millidecibels. The compromises involved in cost reducing the circuit and using linear potentiometers lead to some slight non-linearities. The ¼ and ¾ positions are not exactly 10 and − 10 dB, meaning that the potentiometers are most sensitive towards the end of their travel. This may be preferable to the listener, giving a fine adjustment near the middle of the potentiometers, and more rapid adjustment near the extreme positions. The center frequency shifts slightly, but this should be inaudible. The frequencies nearer the mid-range are adjusted more rapidly than the frequency extremes, which also may be more desirable to the listener. A tone control is not a precision audio circuit, and therefore the listener may prefer these compromises.

D.10 Curve-Fitting Filters

Analog designers are often asked to design low-pass and high-pass filter stages for maximum rejection of frequencies that are out of band. This is not always the case, however. Sometimes, the designer is asked to design a circuit that will conform to a specified frequency response curve. This can be a challenging task, particularly if all the designer knows is that a single-pole filter rolls off 20 dB per decade, and a double-pole 40 dB per decade. How does the designer implement a different roll-off? The obvious place to begin would be with different filter responses such as Chebyshev and Bessel, but these responses have undesirable effects in the passband. So the options are back to the Butterworth response.

It is not possible to get more out of a filter than it is designed to produce. A single pole will give no more than 20 dB per decade, and cannot be increased or decreased. More roll-off demands a double-pole filter with 40 dB per decade. If a designer cannot live with these set values, allow filters at closely spaced frequencies to overlap.

One popular curve-fitting application is the Recording Industry Association of America (RIAA) equalization of Figure D.11, which compensates for equalization applied to vinyl record albums during manufacture. Many newer pieces of audio gear have omitted the RIAA equalization circuit completely, assuming that the majority of users will not desire the function. In spite of the enormous popularity of audio compact discs (CDs), there is still a dedicated group of audiophiles who have a large library of record albums, including titles that are not available on CDs or are out of print.

RIAA has three breakpoints:

- 17 dB from 20 to 50 Hz
- 0 dB from 500 to 2120 Hz
- − 13.7 dB at 10 kHz.

RIAA equalization curves often include another breakpoint at 10 Hz to limit low-frequency "rumble" effects that could resonate with the turntable's tonearm. The

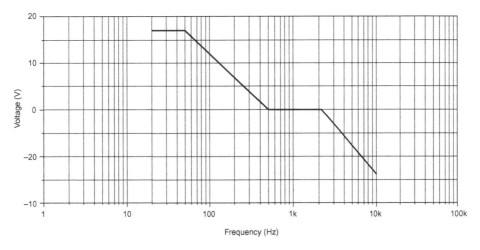

Figure D.11
RIAA Equalization Curve

standard input impedance in the circuits shown here is 47 kΩ. This impedance makes a convenient place to inject DC offset into single-supply circuits, so it is isolated from the phono cartridge by an input capacitance. The phono cartridge output is assumed to be 12 mV.

Application circuits were evaluated from many sources in print and on the web. Many of these did not work at all, did not easily translate to single-supply operation, or deviated markedly from the RIAA specification. Many circuits have been proposed for this function; in fact, competitions have been held to propose the best. Figure D.12 shows a good example.

This circuit topology is very flexible; most of the RIAA breakpoints are independently adjustable:

- R_1 and C_1 set the low-frequency response.
- U_{1A}, R_2, and R_3 control the overall gain of the circuit.
- R_4 and R_5 control the low-frequency gain.
- R_5 and C_2 control the 50 Hz low-frequency breakpoint.
- C_3, C_4, C_5, R_6, R_7, and the op amp form a 500 Hz highpass filter that reverses the effect of the 50 Hz low-pass filter and flattens the response through 1 kHz until the 2120 Hz low-pass filter begins to affect the response.
- R_8, R_9, R_{10}, C_6, C_7, and the op amp form the 2120 Hz low-pass filter; the input resistor has been split into a summing resistor.

The overall response of the filter is shown in Figure D.13.

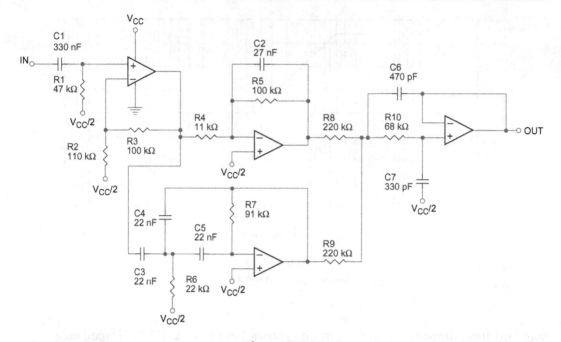

Figure D.12
Equalization Preamplifier

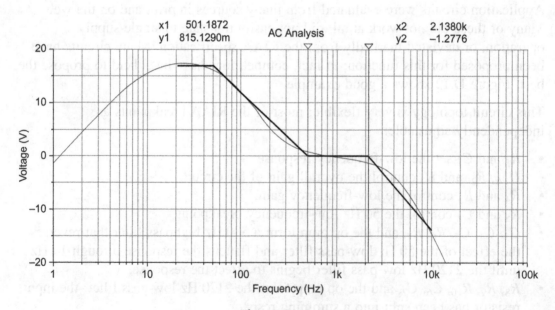

Figure D.13
RIAA Response

The 500 Hz response is above the ideal curve by 0.8 dB, and the 2120 Hz response is below the ideal curve by -1.3 dB. This circuit is about the best that can be made without many more op amps and complex design techniques. It should produce very aesthetically pleasing sound reproduction.

Index

Printed in the United States
By Bookmasters